# Populations and Ecosystems

Developed at
The Lawrence Hall of Science,
University of California, Berkeley
Published and distributed by
Delta Education,
a member of the School Specialty Family

1465671
978-1-62571-176-2
Printing 2 – 8/2016
Webcrafters, Madison, WI

# Table of Contents

# Observations and Inferences

Look at the picture above. What do you observe? Make a list of your **observations** in your science notebook.

Observations provide factual information through the five senses: sight, touch, smell, hearing, and taste. Observations can be qualitative (describing a quality), using adjectives such as *small* or *shiny*, or quantitative (describing a quantity), using numbers such as *3* centimeters (cm) or *20* seeds.

**Inferences** are explanations or assumptions that people make based on their knowledge, experiences, or opinions. Explanations and assumptions are different from observations. Two people could observe the same thing but make different

inferences. "The grass is wet this morning" is an observation. If you didn't observe the grass getting wet, you might infer "It rained last night" or "The sprinklers came on this morning."

Sometimes our inferences are limited by our ability to get the facts. For example, when we observe a milkweed bug heading toward the food source in its **habitat**, we might infer that the bug feels hungry. But we can't really know how it feels. We can only know what we observe, which is the direction in which it is moving.

Let's make some observations again, this time focusing on the difference between observation and inference. Analyze the picture of the bonobos on the next page.

In your notebook, make a T-chart with one column labeled "Observations" and the other labeled "Inferences." Put each of the statements below in one of these columns.

1. There are two **individuals** of the same kind of animal.
2. These animals are in the wild.
3. One animal is mad at the other.
4. Both animals have open mouths and are looking toward each other.
5. Both animals are in the same tree.
6. The smaller animal is holding onto a higher branch.
7. The smaller animal is the child of the larger animal.
8. The smaller animal is scared.

Humans are very social, emotional animals. We have a great capacity for empathy—the ability to understand and share the feelings of another. Sometimes we make inferences that we think explain another animal's actions, but we have no way to check these inferences. For example, we might assume that the larger bonobo is mad at the smaller one. However, there could be another explanation for their actions that we don't know about. The assumption that other animals think like humans is called anthropomorphism.

Scientific thinking starts with detailed, accurate observations. Then, scientists can evaluate their data to make inferences that may help them understand the phenomena they have observed. Just like scientists, when we make inferences, we need to be clear that they are not observations.

Look in your notebook at your first observations of the milkweed bugs. Were any of your observations actually inferences? If so, then mark them with an I. Is it possible to replace them with observation statements?

Each time you make an observation throughout this course, think about whether you are actually making an inference. Challenge yourself to stick to the facts!

What are your observations of the bonobos?

# Milkweed Bugs

Milkweed bugs are easily recognized as insects. They have the same structures as most other insects: six legs, three body parts (head, thorax, and abdomen), and two antennae.

Milkweed bugs are true bugs because they do not have mouths for biting and chewing **food**. Instead they have a **proboscis**, which is a tube-like beak for sucking fluids.

In nature, the milkweed bug uses its proboscis to pierce and suck nutrients from the seeds of the milkweed plant. The bugs in your classroom, however, have been bred to feed exclusively on raw, shelled sunflower seeds. Your classroom milkweed bugs insert their long beaks into sunflower seeds to suck out the oils and other nutrients.

## How Do Milkweed Bugs Grow?

Milkweed bugs start life as tiny eggs. When the eggs hatch about a week after being laid, they are not much bigger than the period at the end of this sentence. If you look at a newly hatched milkweed bug under a microscope, you will see that it is indeed a tiny insect, with six legs, three body parts, and two antennae. You will also see that it has a tough outer covering called an **exoskeleton** to protect it.

The exoskeleton is not flexible, so the tiny bug cannot grow while the exoskeleton is in place. After a few days, the immature milkweed bug, called a **nymph**, bursts its exoskeleton and sheds it. The nymph is then protected by a new exoskeleton. The new exoskeleton is moist and flexible, and the bug pumps itself up, growing to twice its original size in a matter of minutes. Within a few hours, the exoskeleton hardens, and the larger nymph goes about its business of eating and growing.

About a week after the bugs hatch, brittle little transparent exoskeletons with black squiggly legs start to accumulate in the bottom of the habitat. Shedding the exoskeleton in order to grow is called **molting**. Just after molting, the bug is creamy yellow with bright red legs and antennae. Within a few hours, the body turns darker orange, and the legs and antennae become black again.

The milkweed bug molts five times before it becomes a fully mature adult. With each molt, the body shape changes, and the bug develops more dark body markings. Soon, wings start to form. Each nymphal stage is called an **instar**. The first instar is the newly hatched bug, and the fifth instar is the one just before adulthood.

This gradual maturing of an insect is called **incomplete metamorphosis**. The bug gets

**Milkweed-bug life cycle**

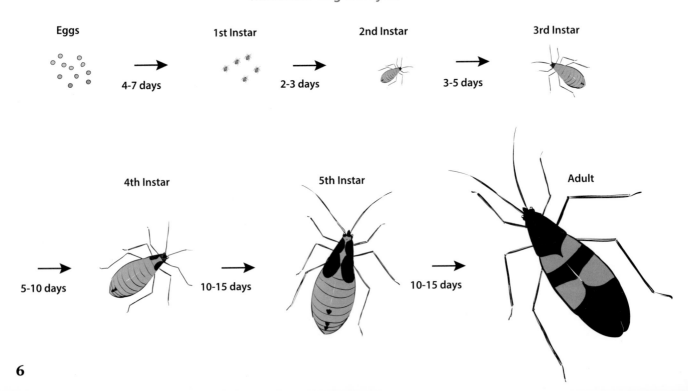

Eggs — 4-7 days — 1st Instar — 2-3 days — 2nd Instar — 3-5 days — 3rd Instar

5-10 days — 4th Instar — 10-15 days — 5th Instar — 10-15 days — Adult

Milkweed bugs mating

steadily bigger and more complete until the last molt reveals the adult. The process from egg to adult takes 5–8 weeks, depending to a large extent on the temperature. A week or more after reaching adulthood, the bugs will mate, and the female will lay eggs. In a room that is a comfortable temperature for humans, the eggs will hatch in about another week, changing from lemon yellow to tangerine orange as they mature. The bugs that hatch will be about half males and half females. The life cycle from egg to egg is about 2 months.

Look at your milkweed-bug observation log. What evidence of bug growth have you observed? Which of the nymphal stages (instars) have you observed?

## What Are the Mating Habits of Milkweed Bugs?

You can easily observe mating, as the two mating bugs remain attached end-to-end for several hours.

It is possible to distinguish the female adults from the male adults by the body markings on the ventral (belly) side of the bugs. The tip of the abdomen is black on both sexes. Next comes a solid orange segment (with tiny black dots at the edges). If the next two segments after the orange segment are solid black bands, it is male.

If the next segment after the orange segment is orange with two large black spots, followed by a solid black band, it is female.

**Female**

Males tend to be smaller than females. When you see a mating pair, observe closely to see if the female is noticeably larger than the male.

A few days after mating, the female starts laying clusters of 20 or more yellow eggs. The clusters are called **clutches**. A

**Male**

female might lay five or more clutches. In nature, the female lays her eggs in a ball of milkweed seed fluff or under a bit of bark for protection. In your classroom habitat, she will usually lay them in the polyester wool.

After mating and laying eggs, adult milkweed bugs might live another 2 months in a kind of retirement. The life span of the milkweed bug in a sheltered habitat bag in a classroom is about 4 months. In the wild, they probably don't live that long.

In captivity, a milkweed-bug **population** will continue to reproduce one generation after another. If the 2-month life cycle continues, six generations can be produced in 1 year! That's potentially a lot of milkweed bugs.

In the wild, however, milkweed bugs stop reproducing in the fall when the weather gets cold and the milkweed plants die. The adult bugs that have not reproduced find shelter where they can hibernate. Even though they have natural antifreeze in their bodies, winter takes its toll on the population. But life is durable, and at least a few bugs always live until spring. And when a male and a female survivor meet on a fine spring day, they naturally continue the process of building up the population.

**Look at your milkweed-bug observation log. When did you first observe evidence of mating in your habitat? What evidence have you observed?**

## Oncopeltus fasciatus

For purposes of classification, all life on Earth is divided into three domains—Bacteria, Archaea, and Eukaryota. Within Eukaryota, scientists have classified **organisms** into kingdoms. The kingdoms are further divided into major groups called phyla. Each phylum is divided into classes. Each class is divided into orders. Each order is divided into families. Each family is divided into genera, and each genus into **species**. A species is a basic category or a kind of organism. A species consists of individuals that are similar in structure and that can breed to produce offspring.

The milkweed bug studied in this course is a member of the phylum Arthropoda, the class Insecta, the order Hemiptera (true bugs), the family Lygaeidae (seed bugs), the genus *Oncopeltus*, and the species *fasciatus*. Its common name is large milkweed bug, and its scientific name is *Oncopeltus fasciatus*. A close relative of the large milkweed bug is *Oncopeltus sexmaculatus*, the six-spotted milkweed bug (same genus, different species). Another more distant relative of these two bugs is *Lygaeus kalmii*, the small milkweed bug (same family but different genus and species).

# Life in a Community

You live in a **community**. But it might not be the community you think you live in. Most people grow up with the idea that their community is the homes, businesses, roads, institutions (schools, hospitals, government offices, police station, library, etc.), and people they interact with regularly. A community is people meeting, working, planning, and living together.

That's one meaning of community. But that's not what *community* means to an ecologist. From a scientific point of view, the community you live in comprises the populations of plants, animals (including humans), and other living organisms that live and interact in an area. Furthermore, a community includes only the populations living in an area, not the place where they live. Your ecological community

might include one or more populations of rodents, several populations of trees, many populations of grasses, hundreds of populations of insects, and countless populations of microscopic organisms such as algae, fungi, and bacteria. Now that's your community.

A community is described by the organisms living in it; therefore, no two are exactly the same. The community of organisms living on an island in Lake Superior is similar to, but not the same as, the community living on the shore of Lake Superior near Marquette, Michigan. Both of those communities, however, are very different from the community of organisms living on the coral reef just off Key West, Florida.

| Island | Lakeshore | Coral Reef |
|---|---|---|
| Sparrow | Sparrow | Algae |
| Crow | Crow | Sea grass |
| Osprey | Osprey | Sponge |
| Mosquito | Mosquito | Coral |
| Black fly | Black fly | Lobster |
| Badger | Deer | Moray eel |
| Skunk | Badger | Angelfish |
| Squirrel | Fox | Gobi |
| Mouse | Skunk | Nudibranch |
| Pine | Mouse | Snail |
| Spruce | Moose | Sea star |
| Oak | Pine | Octopus |
| Beech | Spruce | Sea urchin |
| Blueberry | Oak | Shark |
| Moss | Beech | Porpoise |
| Grass | Blueberry | Shrimp |
|  | Grass | Barracuda |

Another factor that defines a community is the interactions among organisms. In a community, every organism's life is connected to every other organism's life in some way.

Some interactions are obvious. When a robin eats a worm, the robin is fed and the population of worms decreases by one. Other interactions are not as easily observed. You might overlook the importance of an ash tree as a safe place for a robin to build a nest. Without the protection provided by the tree, however, the robin population might decrease.

When the robin dies and its body falls to the forest floor, populations of bacteria and fungi consume the remains, increasing their populations and recycling the minerals from the robin's body into the **environment**. The

A robin eating a worm

ash tree benefits from the mineral nutrients returned to the soil, helping it survive. A healthier ash tree is more likely to grow larger and produce seeds to reproduce new ash trees. The robin's mineral remains, recycled into the soil, nourish the tree, which in turn provides more nesting sites for the next generation of robins. The death of one robin affects not just that robin, but could benefit the entire population of robins.

This is just one peek into the complexity of a community. Interactions among organisms, together with their **abiotic** (nonliving) surroundings, are called an **ecosystem**.

What organisms do you interact with in your community? Which organisms do you eat, and which ones eat you? Which organisms compete with you for food, and which ones provide shelter or comfort?

## Think Questions

1. List the organisms in your community.

2. What are some of the organisms in your community that you cannot see?

Arctic National
Wildlife Refuge

Yellowstone
National Park

Mono Lake
Monterey Bay National
Marine Sanctuary

Delaware Water Gap
National Recreation Area

Tallgrass Prairie
National Preserve

Monongahela
National Forest

Sonoran Desert
National Monument

Everglades
National Park

Florida Keys National
Marine Sanctuary

El Yunque Caribbean
National Forest

# Ecoscenario Introductions

Can you imagine yourself snorkeling off the shore of a beach, or snowshoeing through a landscape of geysers and hot boiling pools? You can do both of these activities, and many more, all within the United States.

Every place on Earth has certain conditions that define it. Some of the most important are conditions of water, air, soil, and local geology. We call these *environmental conditions*. Because the United States extends from the Arctic to the tropics, there are amazingly different environmental conditions throughout the country. The United States may be the most environmentally diverse country in the world.

Because of its environmental diversity, the United States has a tremendous diversity of

organisms. Something is living wherever you look—on the land, in the water, on mountain peaks, on desert sands, and everywhere else.

An ecosystem is defined by the physical environment and the organisms that live in that environment. Ecosystems that have similar environments and organisms are called **biomes**.

In this course, you are going to learn about 11 different ecosystems. They are located in national parks, monuments, sanctuaries, and preserves. The ecosystems and their issues are presented in ecosystem stories that FOSS calls ecoscenarios. Their locations are indicated on the map, and brief introductions to ten of the ecosystems are presented in this article. An 11th ecosystem will be introduced in the next investigation.

A coyote can survive in many ecosystems.

For each ecoscenario, you will learn about the key organisms that populate that ecosystem. Some organisms have a flexible lifestyle. They can survive in a number of ecosystems. For example, coyotes can live in a wide range of temperatures, use many food sources, and are fast runners, secretive, and social. These characteristics allow them to live in a variety of ecosystems, including forests, grasslands, and deserts.

Saguaro cacti, on the other hand, are specialized for one set of physical conditions. They require hot, arid (dry) conditions, with little rain and a bit of shade during their first decade or so of life. They can survive only in the conditions found in the Sonoran Desert.

One species of organism is unique. *Homo sapiens*, or humans, can live in all the ecoscenario locations. If the environment is unsuitable for humans, such as in extreme cold or ocean depths, we craft suitable environments for ourselves in these locations. Warm clothes and housing, imported food and water, and a self-contained air supply make it possible for us to live in the Arctic and underwater.

While many humans no longer live directly off the land, all of us depend on the environment around us. At first, you might not think you get much from the ecosystem you live in. But think about some of your basic human needs, such as food and water. For most Americans, drinking water comes out of a faucet, and much of the food we eat comes from a store. But how did the water get into the pipes that lead to your faucet? Where did the food in the store come from? From the environment.

Remember thinking about snorkeling and snowshoeing? Recreation is another benefit provided by the environment. We call benefits that we obtain from the environment **ecosystem services**. Tourism, protection from flooding, and soil formation are also ecosystem services. Can you think of other important services you get from the environment?

Humans benefit from and affect every ecosystem on Earth. Human actions in ecosystems raise issues about the well-being of the ecosystem. The benefits for humans must be measured against the consequences of these actions on the ecosystems. Each ecoscenario tells a story about the ecosystem that defines it, the ecosystem services it provides, and how humans affect it. Ecosystem services can be categorized as **provisioning**, **regulating**, **cultural**, and **supporting**.

Recreation such as snorkeling is an ecosystem service.

# 1. Arctic National Wildlife Refuge

The Arctic National Wildlife Refuge is in the northeastern corner of Alaska. It is one of the most pristine, undisturbed places on Earth. Most of the refuge's wildlife lives in the 6,070 square kilometers (km²) coastal plain.

### Biome: Arctic Tundra

Arctic tundra forms a band around the North Pole. It is land that is permanently frozen a few centimeters (cm) below the surface. Lichens, sedges (a kind of grass-like plant), and low shrubs cover this frozen landscape. Tundra supports large numbers of seasonal insects (mosquitoes, black flies), migratory birds (ducks, geese), resident birds (willow ptarmigans, snowy owls), large mobile mammals (caribou,

A snowy owl

wolves, brown bears), and smaller mammals (voles, lemmings, snowshoe hares, minks).

Because it is located above the Arctic Circle, Arctic National Wildlife Refuge has long, cool summer days and long, dark, freezing winter nights. This frigid climate shapes the terrain and organisms found there.

A caribou

A brown trout

## 2. Delaware Water Gap National Recreation Area

The Delaware Water Gap National Recreation Area is a 64.3-kilometer (km) stretch of the Delaware River running between New Jersey and Pennsylvania and down through Delaware and Maryland. The Delaware River is the largest free-flowing river in the eastern United States—no dams block the river's flow to the ocean.

### Biome: Freshwater

The freshwater biome includes rivers, streams, lakes, and ponds. Habitats along the edges of streams and rivers are often populated by trees, shrubs, grasses, and vines. The cool, clear, turbulent water has a high concentration of oxygen, which is essential for many **aquatic** organisms.

Freshwater algae are tiny microscopic organisms that produce food for many other organisms in the ecosystem. In addition, many **terrestrial** plants grow along the water's edge. The Delaware River shore is lush with trees (hemlock, maple, oak, hickory), shrubs (mountain laurel, rhododendron), and wildflowers.

Freshwater biomes support a rich diversity of animal life in and along the water's edge, including fish (brook trout, catfish, yellow perch, largemouth bass), mollusks (freshwater mussels, aquatic snails), insects (butterflies, mayflies, bees, ants, crickets, cicadas, flies), and amphibians (frogs, toads, salamanders). Many birds (ducks, bald eagles, swallows, hawks), reptiles (turtles, snakes, lizards), and mammals (squirrels, shrews, wood rats, deer, opossums, beavers, skunks, raccoons) also depend on living close to the water's edge.

Winters in the Delaware Water Gap are cold, and most of the river freezes over. Summers are warm and humid, with thunderstorms and dense fog. The flow of the river is generally smooth and quick. Areas of shallow streams and quiet pools are excellent locations for observing animal life.

A beaver

A bat

## 3. El Yunque National Forest

Puerto Rico is an island in the Caribbean Sea southeast of Florida. El Yunque National Forest is on the eastern end of Puerto Rico. The mountaintops in this area experience warm rainfall all year, producing a lush tropical rain forest.

### Biome: Tropical Rain Forest

Tropical rain forest ecosystems have dense vegetation made up of a wide variety of plants (huge trees, vines, shrubs, low-growing plants). Rain forests support large populations of insects (beetles, moths, ants, termites, flies, wasps, butterflies), large diversity in birds (hummingbirds, orioles, parrots, todies, swifts, warblers, hawks), reptiles and amphibians (snakes, lizards, frogs), and small populations of mammals (bats, rats, cats).

Puerto Rico, between the Tropic of Cancer and the equator, has a tropical climate. Rainfall is consistent and heavy throughout the year. Warm temperatures and daylight hours are also fairly constant year-round. The top of El Yunque peak is shrouded in clouds and has slightly cooler temperatures than the lower forests.

A tropical frog

# 4. Everglades National Park

Everglades National Park is a subtropical wilderness on the southern tip of Florida. The Everglades is a slow-moving, shallow, wide river dominated by saw grass. This river starts at Lake Okeechobee and flows south to the Atlantic Ocean, Florida Bay, and the Gulf of Mexico.

## Biome: Wetland

Wetlands are covered in shallow water most of the year. They occur most often in temperate and subtropical regions and are usually very productive, due to rapid growth of aquatic plants. The communities of plants vary throughout the ecosystem, depending on how deep and salty the water is in a particular location, but all include plants that thrive in or near water (saw grass, mangroves, pine trees).

Wetlands can be densely populated with insects (dragonflies, beetles, mosquitoes, flies), fish (catfish, mosquito fish, bass, perch), reptiles (cottonmouth snakes, turtles,

An American alligator

American alligators, lizards), amphibians (frogs, toads), and birds (anhingas, herons, kingfishers, ducks, thrushes, sandpipers). Wetlands are important nurseries for many fish and other aquatic species. Wetlands are also essential feeding grounds for migratory birds.

An anhinga

The Everglades experiences distinct wet and dry seasons. The wet season is warm, and humidity can be 90 percent or more.

**16**

A lemon shark

## 5. Florida Keys National Marine Sanctuary

The Florida Keys National Marine Sanctuary covers 9,600 km² off the southern tip of Florida. A marine sanctuary is like a national park in the ocean. The reefs in the sanctuary form the third largest system of coral reefs in the world.

### Biome: Coral Reef

Coral reefs develop only in warm, shallow ocean water. Warm seas in the tropics support moderate populations of **phytoplankton** (tiny organisms that carry out **photosynthesis**) and large populations of diverse corals. Corals are tiny soft-bodied animals related to jellyfish. Corals receive food from photosynthetic algae, called zooxanthellae, that live in their bodies. These organisms construct protective living chambers from calcium carbonate, forming the structure you see on the outside of coral. Billions upon billions of these structures produce massive, complex reefs around the world.

Coral reefs support extremely diverse populations of invertebrates (clams, crabs, shrimp, lobsters, octopuses, snails, sea stars, anemones, urchins, sponges, worms) and fish (rays, groupers, gobies, angelfish, wrasses, eels, surgeonfish, parrot fish, sharks).

A moray eel

The Florida Keys lies in a tropical ocean with a tropical climate. Temperatures throughout the year range between 27 degrees Celsius (°C) in summer and 20–24°C in winter. The water receives direct sunlight that penetrates and reflects off the white bottom. This solar energy keeps the water warm year-round.

## 6. Monongahela National Forest

Monongahela National Forest is in the Allegheny Mountains of West Virginia. This national forest covers over 3,678 km², making it the fourth largest national forest in the northeast United States. The landscape is rugged, with spectacular views of exposed rocks, spring wildflowers, and colorful fall leaves.

### Biome: Deciduous Forest

Deciduous forests, also called hardwood forests, are located in temperate regions that experience moderate rainfall, have cold winters, and have fairly rich soils. Deciduous trees (maples, oaks, ashes, beeches, birches, hickories, poplars), shrubs, grasses, and low plants dominate this biome. Hardwood forests support large seasonal populations of insects (gnats, black flies, bugs, beetles, moths, butterflies, ants, bees, wasps, termites, crickets) and birds (warblers, thrushes, hawks, owls, sparrows, swallows, swifts, turkeys, grouse, woodpeckers), and modest populations of reptiles (snakes, turtles), amphibians (salamanders, frogs, toads), and mammals (bears, deer, moose, coyotes, foxes, bobcats, skunks, raccoons, woodchucks, squirrels, mice, voles, gophers).

Monongahela has warm summers and cold winters. There is no distinct rainy season, but the shape and position of the mountains cause the western side of the mountains to receive more rainfall than the eastern side.

A gopher

A turkey

## 7. Monterey Bay National Marine Sanctuary

Monterey Bay National Marine Sanctuary, located off the coast of California, is one of 13 national marine sanctuaries. The central coast of California supports a unique ecosystem known as the kelp forest. Kelp grows from the rocky seabed to the ocean surface.

**Biome: Kelp Forest**

Kelp, a brown algae that carries out photosynthesis, grows underwater in large groups called forests. There are two dominant kinds of kelp in Monterey Bay. The more common species is giant kelp. Giant kelp can grow as much as 0.6 meters (m) a day when it gets enough light. The other species is bull kelp. It grows mostly in exposed areas with plenty of water motion.

Giant kelp with fish

Kelp forests occur in cold, shallow ocean water, populated by dense growths of kelp of various species. Kelp forests support large populations of many classes of invertebrates (sea urchins, lobsters, crabs, octopuses, sea stars, anemones, abalones, snails), fish (sheep heads, garibaldi, sharks, sea bass, rockfish), and marine mammals (seals, sea otters).

The climate in the Monterey Bay National Marine Sanctuary is relatively mild, with warm summers and cool winters. The waters in the sanctuary are influenced by cold Arctic and warm tropical waters. El Niño, a climate pattern that occurs every several years, can warm the water in the sanctuary by several degrees.

A sea otter

## 8. Sonoran Desert National Monument

The Sonoran Desert is located in Arizona, part of California, and a bit of Mexico. Sonoran Desert National Monument protects 1,970 km² of the desert south of Phoenix, Arizona. This hot, dry desert is the most lush and diverse of the North American deserts.

### Biome: Desert

Deserts are characterized by rocky, sandy land, little precipitation, and high temperatures. They are populated by large cacti (saguaro, prickly pear, cholla) and tough, drought-resistant trees and shrubs (mesquite, creosote bush). The Sonoran Desert supports a wide variety of insects (ants, beetles, moths), reptiles (rattlesnakes, lizards, desert tortoises),

A saguaro cactus

birds (red-tailed hawks, cactus wrens, roadrunners, quail), and mammals (coyotes, bats, jackrabbits, kangaroo rats).

Summers in the Sonoran Desert are hot, with temperatures 38 to 46 degrees Celsius (°C). Winters are mild during the day, but can drop to 4°C at night. The rainy season stretches from July to September, yet this desert region receives only about 25 cm of rainfall a year.

A rattlesnake

## 9. Tallgrass Prairie National Preserve

Covering 45 km$^2$ in Kansas, the Tallgrass Prairie National Preserve was created in 1996 as a partnership between the Nature Conservancy and the National Park Service. The grassland prairie is a remnant of the prehistoric prairie that used to cover much of the United States.

### Biome: Temperate Grassland

Temperate grasslands, also known as prairies, are characterized by great expanses of flat or rolling land covered by fairly rich soil. Grasslands receive moderate amounts of precipitation. This can happen in seasonal downpours followed by periods of drought. They often experience hot summers and cool to cold winters. Grasslands are dominated by a number of varieties of grasses and other small, low-growing plants.

Grasslands support modest resident populations of small- and medium-sized mammals (ground squirrels, prairie dogs, ferrets, foxes,

A prairie dog

coyotes) and birds (hawks, prairie chickens, meadowlarks), and large seasonal populations of insects (grasshoppers, bees, beetles, butterflies, moths, locusts, aphids), as well as migratory populations of birds (cranes, bluebirds, sparrows, swallows, horned larks) and large mammals (bison, pronghorn antelope).

The climate of Tallgrass Prairie National Preserve is semiarid (dry), with warm summers and cold, dry winters. Prevailing winds during much of the year can make this grassland even drier.

American bison

## 10. Yellowstone National Park

In 1872, the US Congress established Yellowstone National Park, the first park of its kind in the world. Most of the park is in Wyoming, with small areas in Idaho and Montana. The park includes areas of forest, freshwater lakes and streams, grasslands, and unusual colorful thermal-pond ecosystems.

### Biome: Taiga

Taiga, also known as boreal forest, is land in the northern latitudes, often in mountains, that receives moderate rain, experiences long cold winters and shorter warm summers, and is populated by dense stands of evergreen trees (pines, firs, spruces). Taiga forests support large populations of insects (flies, beetles, moths, wasps, crickets, butterflies), birds (warblers, thrushes, woodpeckers, sparrows, hawks, ravens, swallows), small mammals (squirrels, foxes, weasels, otters, raccoons, rabbits, mice), and large mammals (deer, bears, coyotes, wolves, moose, elk).

Yellowstone experiences long, cold winters and short, mild summers. Snowfall begins in autumn, and by November most roads in the park are closed due to deep snow. Snow typically reaches a depth of 3.8 m over the winter. Snow is also common in the spring and can fall in the upper elevations throughout the summer.

Red foxes

A grizzly bear

# Defining a Biome

At this point, you have learned the definition of an ecosystem. You know it is a combination of **biotic** and abiotic factors.

You also learned that ecologists group similar ecosystems into larger regions. These regions are called biomes, and they have similar climate conditions and communities, even though they might not be near each other on Earth. Biomes can be terrestrial or aquatic. Biomes are described in terms of abiotic factors like climate and soil type, and in terms of biotic factors like plant and animal populations that can survive in these environmental conditions. All the biomes on Earth combine to form the **biosphere**.

**Climate.** Climate is one of the most important environmental conditions that define a biome. Weather patterns, including temperature, air pressure, wind, and precipitation, have been recorded for decades.

The patterns of weather conditions over long periods of time make up the climate.

Locations have different climates partly because of their latitude, which means how far north or south on Earth they are located. Earth can be broken into three main climate zones. The **tropical zone** is closest to the equator (latitude 30° south to 30° north). Both land and water receive sunlight a constant 12 hours a day throughout the year, creating a stable, warm, wet environment that harbors high amounts of life. Most of Earth's landmasses are found in the **temperate zone** (30°–60° N and 30°–60° S). In this zone, moisture and temperature vary with the seasons throughout the year. The coldest year-round temperatures are found in the **polar zone** (60°–90° N and 60°–90° S).

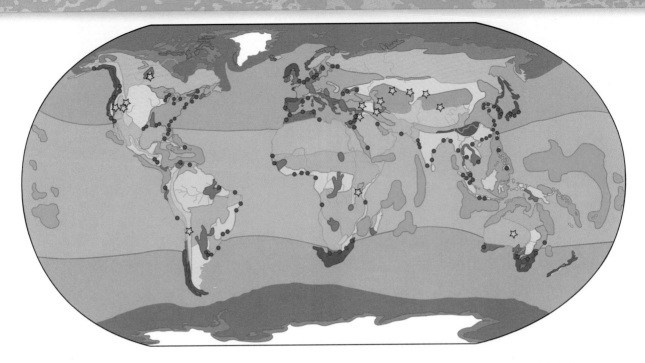

**Abiotic Factors and Aquatic Biomes.** As land-based animals, humans are still learning about the underwater biomes that cover three-fourths of the planet. Half of the ecoscenarios are in aquatic biomes. Abiotic factors that define aquatic biomes include water depth, flow or current, temperature, salinity (salt concentration), and amount of dissolved nutrients.

**Abiotic Factors and Terrestrial Biomes.** Each terrestrial biome's climate is defined mainly by annual patterns of temperature and precipitation. Precipitation can fall in many forms, including rain, sleet, snow, and hail. It is measured in terms of the height of water over the area. For example, in desert biomes, the equivalent of 15 to 25 centimeters (cm) of liquid water falls on the land each year. In contrast, the rain forest biome can receive 125 to 660 cm of liquid water every year. Temperature is often expressed by the average temperature and the temperature range. Temperature range means the difference between the high and low temperatures. For example, many deserts get extremely hot in the day and very cool

at night, with a temperature of 25 degrees Celsius (°C) or more on 1 day, while most tropical rain forests have a temperature range of 14°C or less all year.

Climate is rarely the same throughout an entire area. For example, in the Northern Hemisphere, south-facing sides of hills and mountains receive more sunlight than north-facing sides. You might notice that in a drier biome like taiga, more trees grow in the moister, shaded north-facing sides of mountains. These small differences are called microclimates. For our study, we will focus only on the larger biomes.

**Biotic Factors and Biomes.** Biomes tend to have a defining community of organisms that provides food and shelter for other organisms. For example, trees dominate forest biomes, kelp dominates kelp forests, and coral dominates the coral reef biome.

While areas that share the same biome share similar characteristics, they do not have the exact same biotic communities. These **variations** can be caused by differences in geography, elevation, and local soil conditions.

# An Introduction to Mono Lake

Mono Lake is desolate. Huge expanses of desert stretch out from the lake in three directions, and the mighty Sierra Nevada Mountains rise up on the fourth side. It's freezing cold in the winter and dry and windy in the summer. Sagebrush and desert grasses come right down to the salt-crusted shore. Mono Lake has salt water and it is far saltier than the ocean.

Mono Lake is a unique and ancient ecosystem. It is at least 760,000 years old, making it the oldest lake in the United States. The lake is a tiny leftover puddle, compared to the huge lake complex that covered a bit of California and a large part of northern Nevada and Utah hundreds of thousands of years ago. When the huge lake complex dried up, only Great Salt Lake, Mono Lake, and a handful of other small basins retained water. The minerals dissolved in the lake water concentrated in the small lakes, resulting in high salt concentrations.

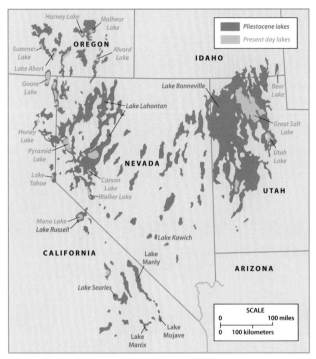

The Pleistocene Lakes

Over the thousands of years that the lake complex existed, it was a place for huge numbers of **migrating** birds to feed and refresh themselves before continuing their journey. As time passed, the lake receded, and the land became dry. The migrating birds came to depend more and more on the remaining small pockets of water. Today, the last remaining bodies of water are essential to the survival of the birds that cross the deserts surrounding Mono Lake.

Even though it looks barren and deserted at first, Mono Lake is one of the most productive ecosystems on the continent. Below the water's surface, a huge population of algae is the base of a **food web**. Massive populations of brine shrimp and brine flies feed on the algae. These crustaceans and insects in turn provide nourishment for populations of birds that come to the Mono Lake basin from as far away as the Arctic Circle and the equator. Millions of birds visit Mono Lake

A California gull feasting on flies

between midsummer and fall, feasting on the abundant food. California gulls and snowy plovers nest on the two protected islands.

When you get down to the lakefront, gray bumpy structures, called **tufa towers**, dominate the Mono Lake landscape. Tufa towers form when calcium-rich water from springs beneath the lake bubbles up through the salt water. Chemical reactions between the calcium and salt create deposits that slowly build tufa towers.

Tufa towers grow only underwater. So why are they sticking up all around the shore of Mono Lake? The answer to this question is found in the recent history of changes in this ecosystem.

Tufa towers

**Abiotic Factors.** Mono Lake, the largest lake (183 square kilometers [km²]) completely in California, sits in a broad basin with no outlet. It is located in an active volcanic area of California. The two islands in the lake, Negit and Paoha, are small volcanic cinder cones. Paoha, the larger island, still spouts steam from vents and has several hot springs.

The most important abiotic factor in Mono Lake is the high salt content of the lake water. Mono Lake contains about 280 million tons of dissolved salts. This makes the water alkaline. Alkaline water is basic (as opposed to acidic). A basic substance you might be familiar with is household bleach, which has a pH of around 12. The water in Mono Lake is not as strong as bleach. The alkali salts raise the pH to about 10, and the water feels slippery between your fingers. (For comparison, pure water is neutral and measures pH 7.)

Rain and snow are critically important to the stability of Mono Lake. Average annual precipitation is around 35 centimeters (cm), which is a very small amount. In addition to rainfall, fresh water enters the lake from the many streams flowing from the Sierra Nevada Mountains, especially during the spring snowmelt. Streams flow into Mono Lake, but no streams flow out. When water in the lake evaporates, the alkali salts remain. Over hundreds of thousands of years, these salts have raised the salinity of the water.

**An aerial view of Mono Lake**

**Winter at Mono Lake**

Fresh water coming into the lake is essential to offset the water lost to evaporation. If less water comes into the lake than the amount of water lost to evaporation, the lake will shrink, but the amount of salt will not. As a result, the lake water gets saltier.

## Life in Mono Lake

Although Mono Lake is home to none of the organisms you might expect to find in a lake, like fish and frogs, it is filled with life. The base of the food web is planktonic algae. During winter and early spring, the microscopic algae reproduce impressively. By March, the lake is "as green as pea soup" from densely concentrated single-celled photosynthetic algae.

In March, countless brine shrimp hatch out of their hard-shelled cysts on the bottom of the lake. Microscopic at first, the baby shrimp feed eagerly on the planktonic algae. After a few weeks, they mature and produce a huge second generation of brine shrimp. In the months of June and July, 4 or 6 trillion mature brine shrimp, each about 1 cm long, fill the lake.

At the same time, the larvae of the brine flies (also known as alkali flies) that have been dormant on the bottom of the lake, become active and start eating the benthic algae that grow on the bottom of the lake. By midsummer, countless millions of the brine flies are skittering across the surface of the lake.

The bounty of shrimp and flies is eaten by huge numbers of birds. Nesting birds consist of 50,000 California gulls (90 percent of California's breeding population and the second largest colony in the world) and 400 snowy plovers (11 percent of the state's breeding population). Migratory birds include 800,000 eared grebes (30 percent of the North American population), 80,000 Wilson's phalaropes (10 percent of the world population), 60,000 rednecked phalaropes (3 percent of the world population), and smaller numbers of 79 other species of waterbirds.

Between April and October, the bird activity is constant. With the departure of the last migrating birds at the end of October, Mono Lake is, for a brief time, quiet. The surface is devoid of life—no flies and no birds. The lake water is clear and still. The brine shrimp ate almost all the algae and then were eaten by the migratory birds. The brine shrimp egg cysts and the dormant larvae of the brine flies lie waiting on the lake bottom. The chill of autumn closes in, and Mono Lake gets cold. Then the cold surface water sinks to the bottom, displacing nutrient-rich water from the bottom of the lake. Nutrients rise to the surface for the 760,000th time, and the few remaining algae start to reproduce. The cycle repeats.

An eared grebe

## Think Questions

1. What would happen if the amount of fresh water entering Mono Lake increased?

2. What would happen if the amount of fresh water entering Mono Lake decreased?

# Biosphere 2: An Experiment in Isolation

One way scientists think about Earth is as a set of interacting subsystems. The **lithosphere** is the rocky, mineral subsystem of the planet. This is the hard part of the planet that we live on. Earthquakes and volcanoes remind us that the lithosphere is restless and dynamic.

Wrapped around Earth is the **atmosphere**, the thin layer of gases that extends about 600 kilometers (km) above Earth's surface. The atmosphere is a source of essential gaseous chemicals, an energy-transfer system, a shield protecting us from extraterrestrial radiation, and an insulator. It is also an important medium for water distribution.

Earth is a water planet. Because of the temperature on Earth, water exists naturally in three states: liquid, gas, and solid. All the water on Earth makes up the **hydrosphere**. The hydrosphere includes the ocean, lakes, rivers, streams, and aquifers. It includes polar icecaps, glaciers, snowpacks, and permafrost. It also includes the water vapor in the air and water drops in clouds, fog, and precipitation.

And finally, creeping, hiding, running, burrowing, flying, slithering, and swimming through, over, under, onto, and into the other three spheres is the biosphere. This is the sum total of all the living organisms on Earth. This incredible assemblage of millions of different kinds of life-forms gives Earth its particular flavor.

All four subsystems can be bundled into one global Earth system that includes the lithosphere, the atmosphere, the hydrosphere, and the biosphere. The interactions between living organisms and the other three subsystems ensure that life is never disconnected from the physical environment.

The ocean is part of the hydrosphere.

Clouds are tiny droplets of liquid water.

**Biosphere 2 in Arizona**

## Biosphere 2

In order to better understand Earth's biosphere, a group of scientists proposed an ambitious plan.

How do you study an entire ecosystem? You can complete an **observational study** or **population study** in the wild. But how could you complete a **controlled experiment** on an ecosystem? A real ecosystem is much larger and more complex than the habitats you can build in class. To find out how carbon dioxide ($CO_2$) affects an ecosystem, you could add $CO_2$ to a field or forest to see how plant growth might change, but the wind would soon blow the $CO_2$ away.

Or you could monitor the weather for a period of 40 or 50 years and look for patterns, but it would be impossible to **control** variables. You might notice a difference in plant growth and health when you compare a wet, cold, rainy year to a hot, dry, sunny year. Is the difference because of the rain, the moisture in the air, the temperature, or the amount of sunlight? It is impossible to say with so many variables changing at once.

To do controlled experiments, you need a box like your terrarium or aquarium that is big enough to hold a whole ecosystem. Then you could control all the variables except the one you are studying. Where can you find such a box? Biosphere 2.

Can human beings live in an airtight building for 2 years with no life support except for the organisms they bring inside with them? Can they set up an ecosystem that provides for their every need and the needs of the other populations around them? Biosphere 2 was built to answer these big questions.

Scientists use models to develop explanations about the natural world. Models can often make it possible to explore aspects of the real world that are too complicated to easily observe. In 1984, construction began on Biosphere 2, intended to be the world's largest model of Earth as a system.

## Design Challenges

In order to create an accurate model, the designers needed to overcome several major challenges. The first design challenge was to choose a site and design a structure that, like Earth, would be an open system for **energy** but a closed system for matter. This meant that while energy enters and leaves, no materials (even gases) could pass through.

The chosen site was in the Sonoran Desert, in Arizona, where the sky is often cloudless. This site maximizes access to sunlight. Sunshine is essential because it is the only energy source to **sustain** the humans and hundreds of other species in the closed system.

Another critically important factor was sealing the giant live-in terrarium from the outside environment. A greenhouse—made of glass, steel, and concrete—was built on top of a solid steel foundation to completely isolate Biosphere 2 from the surrounding earth and atmosphere. The $150 million chamber was ready for business in 1991.

**Biosphere 2 has a glass greenhouse on top of the foundation.**

An aerial view of Biosphere 2

The second design challenge was to create a self-sustaining ecosystem that could support four men and four women for 2 years. Ecologists carefully selected the organisms to populate Biosphere 2. They knew they needed plants, animals, and microorganisms. Not only did they need to provide food for the human team, they also needed to provide life support for the organisms that would produce that food. They needed organisms to refresh the air and dispose of waste materials. The planning was complex and detailed—lives depended on getting it right.

Biosphere 2 covers almost as much area as four football fields. Inside are seven different environments: rain forest, desert, tropical ocean, marsh, savanna, thorn scrub, and a farm within a forest. The farm area was planted 1 year before the mission began, so that food would be available as soon as the terrarium was sealed.

## The First Mission

In September 1991, the door was closed. The eight "Biospherians" and 1,800 other populations were sealed on the inside. In many ways, these humans became Earth astronauts. The challenges they faced were similar to what astronauts face during long-term space travel. The trip to the Moon takes a few days. A trip to Mars might take 1 year. The most efficient way to make such a journey would be in a mini-ecosystem, where everything needed for life recycles.

Compare your classroom minihabitats with Biosphere 2. What similarities do you see? What differences exist between the models?

**Low oxygen levels.** After only 1 day in a sealed atmosphere, the team noticed that $CO_2$ levels were rising quickly and oxygen ($O_2$) levels were dropping. If they couldn't find the source of the problem and solve it, the entire project would fail. Where was the $O_2$ going?

Analysis revealed a clue in the soil in Biosphere 2. The populations of microbes were growing too fast, using too much $O_2$. The scientists reasoned that if the $O_2$ concentration was going down, the $CO_2$ concentration should be going up. But it was not going up as fast as the scientists calculated. It was later discovered that the $CO_2$ was being absorbed by the building's concrete, which was still curing (a hardening process that uses $CO_2$).

Finding the source of the problem did not solve it. After 16 months, the oxygen had fallen from 21 percent (the normal amount at sea level on Earth) to 14 percent (the amount found at 5,000 meters [m] elevation on Earth). $CO_2$ levels were 12 times higher than those in Earth's atmosphere. Ultimately, oxygen was brought in from the outside world, causing widespread criticism but allowing the humans to continue under safer conditions.

**Uninvited guests.** Two uninvited local species, cockroaches and longhorn crazy ants, got into Biosphere 2 and caused havoc. The cockroach population grew out of control. The crazy ants are excellent communicators, which gave them an advantage over the other insect species in Biosphere 2. The crazy ant population increased while most of the other arthropod populations decreased. Not only did the ants put pressure on other organisms in the ecosystem, they clogged vents and chewed on wiring.

The cockroach population increased quickly in Biosphere 2.

**Small successes.** The Biospherians made it through the 2-year mission, but not without some outside assistance and a lot of criticism. Still, a few great insights came from this first attempt to create a model of Earth.

While the team struggled to produce food the first year, they successfully produced more than 80 percent of their food the second year. And the low-calorie, nutrient-rich diet significantly lowered the cholesterol and blood pressure of several Biospherians and strengthened many of their immune systems.

In addition, the airtight seal of the station allowed scientists to measure how $O_2$ and $CO_2$ concentrations changed throughout a day and over a season. The scientists were able to measure how interactions with plants and soil microorganisms caused this change.

**Later development.** A second mission in 1994 lasted 6 months. After the second mission, Biosphere 2 was no longer a live-in facility. It was transformed into a unique ecological research center. In 1995, Columbia University took over management and continued research for 8 years.

After that, many plans were proposed for Biosphere 2, including converting it to a university, a hotel, and a resort. Since 2011, the University of Arizona has owned Biosphere 2. It is a center of research, teaching, and learning about Earth's living systems. Today, Biosphere 2 is the world's largest earth science laboratory.

**Biosphere 2 is the world's largest earth science laboratory.**

# Recent Research on Carbon Dioxide

You have probably heard about humans using energy sources such as gasoline and coal that release $CO_2$ into the air. More $CO_2$ in the atmosphere is linked to global climate change. What will happen if the $CO_2$ level continues to rise? Scientists hope to use Biosphere 2 to gather information to help answer this question.

**Tropical rain forests.** Tropical rain forests are sometimes called the lungs of the planet because they take in so much $CO_2$ and produce so much $O_2$ during photosynthesis. Dense plants and long days of direct sunlight year-round enable rain forest plants to carry out photosynthesis at a higher rate than anywhere else on Earth. Some scientists think rain forests have great potential for controlling the rising $CO_2$ level.

Another subject of interest is how elevated levels of $CO_2$ might impact rainfall. Will there be more rainfall or less?

Inside the rain forest environment in Biosphere 2, scientists are doing experiments to determine how changes in rainfall affect photosynthesis of rain forest plants. They can control the amount of "rainfall" with the overhead sprinkler system. They can pump in more $CO_2$. The temperature can be regulated with huge heating and cooling units. Scientists can change any of these variables one at a time to see what effect each variable has on the plants and the amount of $CO_2$ in the atmosphere.

**A tropical rain forest environment**

**Coral reefs.** Coral reefs have been called the rain forests of the sea because they support so much diversity of life. Coral defines the reef ecosystem. Algae, fish, crabs, and many other kinds of sea organisms live on, in, and around the coral structures.

**A coral reef ecosystem**

A turtle swims around this coral reef.

Marine biologists report that there are fewer and fewer fish and other organisms living in the coral reef ecosystems around the world and that the growth rate of coral has slowed. Warmer oceanic waters have been linked with coral death. Could more $CO_2$ in the water be affecting coral reef ecosystems?

Several scientists are studying the coral reefs in the ocean biome at Biosphere 2. The Biosphere 2 "ocean" holds 2,500,000 liters (L) of water. The depth ranges from 0 m at the beach to 7 m in the deepest part. Scientists have investigated several factors they think might affect the health and survival of the coral.

When they increased the concentration of $CO_2$ dissolved in the water, the health of the coral declined. They found that excess $CO_2$ in the water prevented coral from getting calcium out of the water to build their skeleton. The studies of the coral reefs in Biosphere 2 provide evidence that the changes taking place in the ocean, such as more dissolved $CO_2$, are damaging coral reef ecosystems.

**Important conclusions.** As scientists learn more about ecosystems, two things become very clear. Any change in one part of an ecosystem affects every other part of the ecosystem, often in ways that no one could have anticipated. Also, the more we learn, the more we realize how complex natural ecosystems are and how little we understand about the way they work or what effect human activity has on them.

Why should we care? What difference does it make if rain forests or coral reef organisms are disappearing? That's not where we live.

Well, it is where we live. Earth is small. The atmosphere surrounds the planet, and the ocean washes up on the shores of all the continents. Changes in one ecosystem are communicated to the rest of the world by flowing air and water. Everything is connected. Small changes in global temperature can have a huge effect on weather patterns. Weather distributes water, and water is life.

Maybe you can join the community of people trying to answer some of these tough ecosystem questions. College students from several US universities attend classes at Biosphere 2 for a semester to study environmental problems. There is also a summer program for high school students who are interested in studying environmental issues.

## Think Questions

1. Give two examples of how a change to one variable in an ecosystem can start a chain reaction that affects several other variables.

2. What are some benefits of doing research on ecosystems in Biosphere 2 rather than in the natural ecosystem?

3. What are some limitations of doing research on ecosystems in Biosphere 2 rather than in the natural ecosystem?

4. Why is the increase of $CO_2$ in the atmosphere such an important problem?

## A rough comparison of Earth and the Biosphere 2 original experiment

| System Variables | Earth System | Biosphere 2 |
|---|---|---|
| Location | Earth, solar system | Oracle, Arizona |
| Footprint | 126 billion acres | 3.1 acres |
| Biomes | 13 | 5 |
| Species | Approx. 10–40 million | 1,800 |
| Human population | Over 7 billion | 8 |
| Energy sources | Sun, fossil fuels | Sun, fossil fuels |
| Age of life | 3.5 billion years | 2 years |
| Enclosure | Magnetic field, gravity | Glass and steel frame, gravity |

Fruits like strawberries are mostly water.

# What Does Water Do?

It's a hot spring day, and you just finished gym class outside. As you head back in, you and your classmates swarm to the water fountain, hoping for a sip of cold, refreshing water. Just a few sips make you feel a bit cooler, less tired, and more comfortable. What is it about water that can make a person feel so much better? What is it about thirst that can make a person feel so terrible?

Only food can give your body energy, but your body needs water to function properly. Your blood, which contains a lot of water, carries oxygen to all the cells of your body. You need water to digest your food and get rid of waste, and water is the main ingredient in the sweat that helps cool your body temperature on hot days. Each cell in your body depends on water to function normally.

Life as we know it would not exist without water. More than half of your body weight is water, and you wouldn't be able to survive for more than a few days without consuming water! Any fluid you drink contains water, but plain water or milk are the best choices. Many foods contain water, including fruits and vegetables.

## How Much Is Enough?

Since water is so important, you might wonder if you're getting enough. How much fluid each person needs depends on his or her age, size, level of physical activity, and the surrounding environment. For example, when it is hot (or you heat up by exercising), your body will sweat more to help cool you down. Sweat is a form of water loss from your body. You can lose a lot of water in just an hour of activity on a hot day.

When your body doesn't have enough water, you can become dehydrated. **Dehydration** means the body doesn't have enough water to function normally. The body is used to small changes in the water level. But when the water loss becomes too great, possibly because you forgot to drink more on a hot day, dehydration can make you feel sick.

It is important to drink water on a hot day.

In addition to sweating, another common cause of dehydration in teens is illness. Sick people can lose fluid through vomiting and diarrhea. Sometimes you might not want to eat or drink if your throat hurts or if you keep vomiting. But it is good to remember that most ailments—whether you are sick, stressed, or injured—can be lessened by drinking water.

A common symptom of dehydration is a headache. Other symptoms include feeling dizzy, having a dry mouth, and producing less, darker urine. As dehydration gets worse, a person will start to feel more and more sick as more organs and body systems are affected by the lack of water. In severe cases, a person may need to be hospitalized to recover.

## Preventing Dehydration

The easiest way to avoid dehydration is to drink water often, when you get thirsty on hot days. When you drink is also important. If you're going to sports practice, helping with yard work, or just playing hard, drink water before, during, and after. It is much better to drink small amounts often than to down a lot at once.

Most of the time, your body keeps you properly hydrated without your knowing it. The body can hold on to water when you don't have enough or get rid of it if you have too much. The water you need comes mostly from the food in your diet. The rest comes from drinking a glass of water or some liquid whenever you get thirsty.

Drinking water while exercising keeps your body hydrated.

## Other drink options

People are often tempted by drink options that promise an additional energy boost, quicker rehydration, or make other claims. Most of these drink options contain sugar. When you are quite thirsty and want to gulp down that drink, you may be gulping down a lot more **calories** (food energy is measured in **kilocalories**) than you realize.

Energy drinks and soft drinks are even worse because they can contain caffeine. Caffeine is a substance that is found naturally in the leaves and seeds of many plants like tea, coffee, cacao (used to make chocolate),

Caffeine is found in cacao beans.

and kola (a tree whose fruit gives "cola" its name). Scientists can also produce caffeine artificially, and it can be added to food or drinks.

Caffeine acts as a stimulant in the body. It can make a person feel more awake and alert, but it doesn't contain energy the way food does. People who consume large amounts of caffeine may feel jittery and experience problems sleeping.

When you feel a little run down, try drinking some water before reaching for an energy drink or other product that contains sugar or caffeine. You can also check the nutrition label on a drink product before you consume it, to make sure it does not have a lot of sugar or caffeine.

Review your observations of the classroom habitats. What evidence do you have that all living organisms need water?

# Where Does Food Come From?

I have a sweet tooth for caramel corn, it's one of my favorite snacks. I have a friend who loves chocolate-covered raisins. Those are both popular snack foods. We eat snack foods when we are a little hungry in the afternoon or maybe at a movie or sporting event. At mealtimes, my favorite foods include cereal, sandwiches, pizza, salad, pasta, and enchiladas. With meals and snacks, it seems like we are eating something just about all the time. Why is that?

We eat to live. For that matter, every animal has to eat to live. Food is the source of two essential elements of life: molecular building blocks and energy.

The molecular building blocks are matter. That is, they are made out of **atoms** and are solid "stuff." Building blocks are **molecules**

that come from the nutrients in food. Without these molecular building blocks, you couldn't grow, you couldn't replace hair or skin cells, and your body couldn't regulate body processes or heal when you get injured.

Energy is not matter. It is not made of atoms. Energy cannot be created or destroyed, but it can be transferred. Light energy from the Sun enters an ecosystem and flows through organisms in the form of energy that is stored in the bonds of food molecules. Energy is needed to do all the things that living organisms do. It is needed to move, talk, digest food, keep warm, think, grow, and feel. All organisms release energy back into the environment as heat. Energy is the drive that keeps the machinery of life running.

A pepperoni pizza

Wheat flour and seeds

## Following the Energy Path

Much of Earth's energy comes from the Sun. But humans don't get the energy to run their body directly from the Sun. And how did the energy get into the food that we eat? Most people in the United States get food at grocery stores and restaurants, but where did that food come from? And what organisms stored light energy in the bonds of food molecules that we eat? Let's find out by following the paths that lead to a pepperoni pizza.

The main ingredients in a pepperoni pizza are bread dough, tomato sauce, cheese, and pepperoni sausage. Start with the dough . . . that's a short path. The pizza dough is bread, made from wheat flour. Wheat flour is made by milling or grinding the seeds of the wheat plant, a kind of grass. So, the crust is a part of the pizza that comes fairly directly from a plant source.

Tomato sauce is made by grinding tomatoes into a pulp and cooking them with seasonings. The tomatoes are the fruit of a tomato plant. The seasonings, such as oregano and basil, are leaves of other plants. So, the sauce is a part of the pizza that comes fairly directly from another plant source. The tomato plant, like all other photosynthetic organisms, used sunlight, water, and carbon dioxide ($CO_2$) to create the building blocks it needed to grow, and to store energy. When we eat the fruit of the plant, those building blocks and stored energy of the plant become our food.

Tomato sauce on pizza dough

Cheese is processed from milk, so the nutrients in cheese come from milk. Milk is produced by cows that eat grass. The nutrients in milk come from that grass. So we can trace the energy and nutrients in cheese back to the first organism in the line, a plant.

And what about the pepperoni? Pepperoni is made from the ground-up muscle of cattle and pigs. The animals grew by eating grasses and grain seeds such as corn and millet, which are plant sources.

It looks like each ingredient in a pepperoni pizza had its origin in a plant source. It's plants that capture the energy in sunlight and transfer it into the bonds of food molecules. Plants use photosynthesis to accomplish this. Their cells use sunlight to rearrange water and $CO_2$ molecules into molecules of sugar (food), releasing oxygen ($O_2$). The sugar molecules are used by plants as building blocks to support growth or to store energy.

When an animal eats a plant, those molecules and stored energy are transferred to the animal. If another animal eats that animal, the molecules and stored energy are transferred yet again. So, when you trace any food back to its source, it all starts with plants or plantlike organisms. Always.

Cheese

A cow

Slices of pepperoni

46

Anchovies

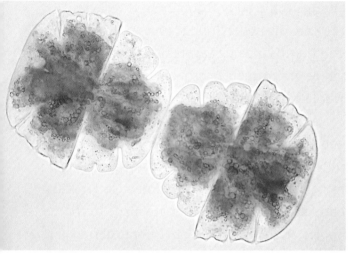

Phytoplankton

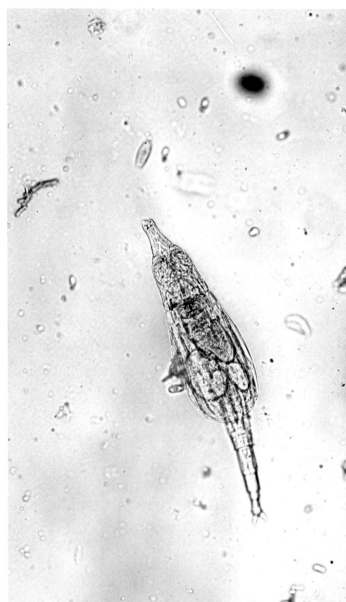

Zooplankton

Did you order anchovies? The case of these little fish is different, but only slightly. Anchovies don't eat plants, as do cows and pigs. Anchovies are animals that eat other animals. The animals they eat are tiny free-swimming critters called **zooplankton**.

But what do the zooplankton eat? They eat even smaller plantlike organisms called phytoplankton. What makes them plantlike is that they can make their own food by photosynthesis, just like plants. So the anchovies eat little animals that eat tiny plants to survive. The anchovies arrive on your pizza because of the food in the microscopic phytoplankton floating in the sea, providing matter and energy.

What's your favorite food? In your notebook, write the name of the food and see if you can record the path of how the Sun's energy got into that food.

# Eating for Energy and Nutrients

Plants and other photosynthetic organisms use the energy in sunlight to turn water and $CO_2$ into food in the form of sugar. The energy is now stored in the bonds of the sugar molecules. Organisms that can't produce food must consume other organisms. At this simple level, it might seem that the whole system is based on sugar. But there's more to the story. Sugars are a form of simple **carbohydrate**. These sugars can be used as building blocks to create other, larger molecules that the plant needs, including starches and cellulose (plant fiber). These larger molecules are called complex carbohydrates. Most plant carbohydrates are so complex that human stomachs are not able to digest them. This is one reason we can't just go out on the lawn and eat grass for lunch.

Simple sugars are the fastest available form of energy, but they don't last long. When a person eats simple sugars, cells are able to access the energy immediately, which also means the cells burn up the energy quickly. This is why you might feel a quick burst of energy after consuming candy or a soft drink, but run out of energy soon after. If the body consumes more simple sugar than it can process, it will store the excess as fat.

To avoid an energy crash, you can eat other carbohydrates that offer a steady amount of energy to the body throughout the day as they are digested. Good sources of foods that will do this include vegetables, fruits, and whole grains such as brown rice, oatmeal, and whole-wheat bread.

Carbohydrates are the foundation of energy stored in food, but your body needs other substances like protein, fat, vitamins, minerals, and water to function well. Proteins and fats are molecules that are made in the bodies of organisms, or can be transferred from one organism to another when an organism gets eaten.

Natural sources of protein

Protein is important for repairing and building cells. People can get plenty of protein through regular healthy meals. Good sources of protein are eggs, dairy, beans, nuts, soy, and animal meats (fish, beef, pork, and poultry).

Fats are foods that offer slow, long-lasting sources of energy. You also need fat in order to digest proteins. Everyone needs a certain amount of fat each day. Some fats, including vegetable oils and nuts, are healthier than others.

Vitamins and minerals are in the food you eat. These nutrients help your cells access energy from the food you eat and keep you from getting sick. You don't get energy from these nutrients, but if you are not getting enough of them, your cells can't effectively use energy. This can make you feel weak or tired!

Two of the more important nutrients for active people are calcium and iron. Calcium helps build strong bones, and iron carries oxygen to muscles. Many teens don't get enough of these minerals. They aren't found in snack foods like soft drinks, potato chips, and candy. Excellent sources of the vitamins and minerals you need include a variety of fresh foods, especially vegetables and fruits.

**Yogurt and blueberries are a healthy snack.**

Leafy, green vegetables such as spinach or kale are a great source of many minerals and vitamins, including iron and calcium. You can also get iron from animal meats. Calcium can be found in some fruits such as blueberries as well as dairy foods like milk, yogurt, and cheese.

Ahh, thinking about nutrients and energy made me hungry. Maybe it's time for a healthy snack!

## Think Questions

1. How does the energy stored in food enter the ecosystem?

2. What was the last food you ate? Draw a diagram to show a possible pathway of energy from when it entered the ecosystem to when you ate it.

Milk is a good source of calcium.

# Energy and Life

All organisms need energy in order to survive. How an organism gets its energy defines its ecological role in the ecosystem. But first, what is energy?

Energy is the ability to do work. Living organisms need energy to perform the basic functions of life, such as growth, reproduction, gas exchange, eliminating waste, getting water and nutrients, and responding to the environment. These processes take place in each and every individual cell, whether the cell is a free-living organism (like an amoeba) or one of millions within a multicellular organism (like a cactus or a human). All cells require energy because all cells are constantly

working. Food is the source of matter and energy for all organisms. But not all organisms eat food. How can some organisms get the benefits of food without eating it? They make food right in their own cells.

## Producers

The Sun is the source of energy for just about all the organisms in every ecosystem on Earth. Some organisms, notably plants, algae, and phytoplankton, capture energy from the Sun using photosynthesis (photo = light; synthesis = putting things together). These organisms are called **producers**.

Producers carry out photosynthesis. During photosynthesis, energy transfers from sunlight into chemical bonds of food molecules (carbohydrates).

Photosynthesis is the chemical process that creates these new molecules. It turns carbon dioxide and water into carbohydrates in the presence of sunlight and **chlorophyll**. Below is a simplified equation that shows how six water molecules ($H_2O$) and six carbon dioxide molecules ($CO_2$) are changed into a carbohydrate molecule (sugar) with energy from light. Note that six molecules of oxygen ($O_2$) are also created.

$$6CO_2 + 6H_2O + \textit{light energy} \longrightarrow C_6H_{12}O_6 + 6O_2$$

carbon    water     energy        sugar     oxygen
dioxide

The sugar molecule becomes food when it is used. Food is built from the atoms that are in $CO_2$ and $H_2O$ and contains potential energy within its chemical bonds. Photosynthetic organisms use the sugar they manufacture as food to get the energy and building blocks they need to live. Plants never eat ham sandwiches, algae never eat enchiladas, and photosynthetic bacteria never eat fruit. They don't have to. Because they make their own food, producers are also referred to as **autotrophs** (auto = self; troph = food).

Producers create **biomass** (bio = living; **mass** = matter). Biomass is the mass of matter produced by organisms in an ecosystem. Every time producers build a new carbohydrate molecule through photosynthesis, their biomass increases. Producers on land include plants such as grasses, shrubs, and trees. Aquatic and

**Giant kelp is a producer.**

marine producers include microscopic phytoplankton, water plants, sea grasses, and seaweeds such as giant kelp.

## Consumers

Producers transform the Sun's energy into potential energy in biomass. But humans are not autotrophs. We can't make our own food. The same is true for all the other animals, the fungi, and most protists and bacteria. All organisms need food, however, so the ones that don't make their own food eat the ones that do. Organisms that eat other organisms are called **heterotrophs** (hetero = other; troph = food). Another name for heterotrophs is **consumers**, because they have to consume or eat other organisms to get energy.

# Aerobic Cellular Respiration

Producers make life possible for consumers because of what they do during photosynthesis: they produce potential energy in the bonds of molecules (carbohydrates) that organisms can use as food. But the producers don't perform photosynthesis as a favor for us. They need the food energy as well. Once the energy is stored in the bonds of molecules, what do producers do with it?

Plants, animals, and nearly every other organism on Earth release the stored energy of carbohydrates in a process called **aerobic cellular respiration**. During this process, the bonds of the carbohydrate molecule are broken and new products are formed ($CO_2$ and $H_2O$). It is through this aerobic respiration process that energy is released. Cells use this energy to do the various kinds of work they need to do. In the process, some energy is lost as heat to the environment.

Here is a simplified equation showing what happens in the cell.

$$C_6H_{12}O_6 + 6O_2 \longrightarrow 6CO_2 + 6H_2O + \text{usable energy}$$

sugar    oxygen    carbon    water    energy
dioxide

Take a close look at the equation for photosynthesis and the equation for cellular respiration. Notice anything interesting? The sugar molecules are on the opposite sides of the equations, and the light energy was turned into usable energy for the plant.

Almost all organisms rely on aerobic cellular respiration to transfer energy from carbohydrates. But only photosynthetic organisms can capture the Sun's energy to create sugars. How do other organisms get sugars? All other organisms must eat photosynthetic organisms, such as plants, or eat other organisms that already ate photosynthetic organisms.

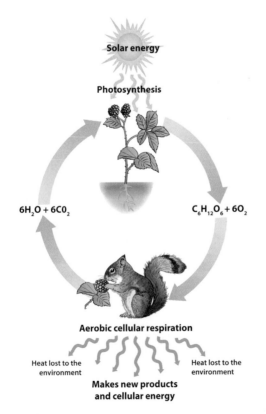

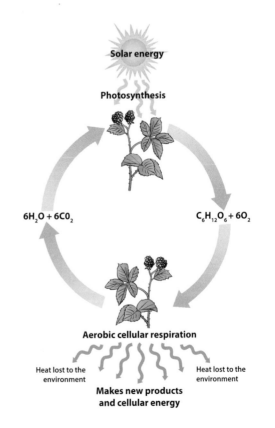

# Wangari Maathai: Being a Hummingbird

You may not have heard about Professor Wangari Maathai (1940–2011), but she won a Nobel Peace Prize and has inspired tens of thousands of people around the world. How? She saw a problem in her local ecosystem and decided she might be able to fix it.

Maathai grew up in Kenya, a country in Africa. By the late 1970s, most of the trees in the forests of Kenya and surrounding areas had been cut down for firewood and buildings. Without trees, the rich soil washed away during rainfalls, making it hard for other plants to survive. The loss of soil and plants also meant that local streams began to dry up, wildlife disappeared, and it became hard to find or grow food.

Maathai noticed that women were suffering most from the hardships of seeking water, firewood, and food for their families. It was an overwhelming situation, but she was inspired by a story she had heard. Here is the story in her own words.

Wangari Maathai

*The story of being a hummingbird is about this huge forest being consumed by a fire. All the animals in the forest come out, and they are transfixed as they watch the forest burning. And they feel very overwhelmed and powerless, except for this little hummingbird. It says, "I am going to do something about the fire." So it flies to the nearest stream, takes a drop water, and puts it on the fire. It goes up and down, up and down, as fast as it can. In the meantime, all the other animals—much bigger animals, like the elephants, with a big trunk that could bring much more water—they are standing there helpless. And they are saying to the hummingbird, "What do you think you can do? You are too little! This fire is too big! Your wings are too little, and your beak is so small, you only can bring a small drop of water at a time!" But as they continue to discourage it, it turns to them without wasting any time, and tells them, "I am doing the best I can."*

**A hummingbird**

Maathai decided to be the hummingbird and do the best she could to help her fellow Kenyans. She knew how important a strong community of producers, such as the trees that had been cut down, was for a healthy ecosystem. So she started a small organization that educated women about basic ecology and helped them learn how to plant and care for trees.

The movement (called the Green Belt Movement) was a great success, but life was hard on Maathai. She was the first woman in East and Central Africa to earn a doctorate degree, and at the time it was rare for women to be so outspoken. Her husband divorced her, she battled public criticism, and she was arrested and beaten several times by the police for her political activism. Still, she continued to work tirelessly toward her goal, just like the hummingbird in her story.

Today, the Green Belt Movement has planted more than 30 million trees in Kenya and other African countries, reshaping the local ecosystems back into healthy forests. As the movement grew, it expanded to help educate and empower women across Africa to understand their local environmental problems, take action to seek solutions, and become leaders in their communities.

In 2004, Maathai, known by many as Mama Miti, or "Mother of Trees," became the first African woman and environmentalist to win the prestigious Nobel Peace Prize. During her acceptance speech, she said, "I would like to call on young people to commit themselves to activities that contribute toward achieving their long-term dreams. You are a gift to your communities and indeed the world. You are the hope and our future."

## Think Questions

1. Can one hummingbird make a difference? Explain your thinking.

2. How could protecting the producers of an ecosystem affect the entire ecosystem? Use specific examples from the article.

3. When presenting Wangari Maathai with the Nobel Peace Prize, the committee explained that she "thinks globally and acts locally." Can you give an example of how you could think globally and act locally?

**Trees in Africa**

# Rachel Carson and the Silent Spring

During World War II, government researchers began manufacturing chemicals called pesticides. The pesticides were used to help kill insects that could transmit diseases to people, including military troops. After the war, farmers began to use these chemicals to kill insects, weeds, and small animals that can destroy food crops. One of the main pesticides used was DDT.

Rachel Carson (1907–1964), a biologist and author, was becoming concerned about how pesticides were affecting the environment. Her books about marine life were very popular at the time. But her interests were about to stray very far from the sea.

Huge quantities of DDT were being sprayed from the air onto crops and even on playgrounds during mosquito season. Carson

Carson

wondered what effects DDT was having on people and the environment. She decided to research the topic and write a book to help people understand.

Carson found that in the case of DDT, the excess chemical entered the ground and water system. From there, it could enter the **food chain** through earthworms and aquatic insects. As birds and fish ate the contaminated worms and insects, they became more and more contaminated. Birds of prey that ate these birds and fish would accumulate even more toxin in their bodies. This is called **bioaccumulation**. As animals in the food chain eat contaminated organisms, the level of contamination in their bodies rises. Organisms at the top of the food chain, such as humans or polar bears, can accumulate the highest concentrations.

During the 1950s, many bird populations dramatically declined because high levels of DDT in their bodies made their eggs so fragile that they broke in the nest. To many people, the most alarming decline was in the bald eagle population. The bald eagle is the national bird and the national animal of the United States. By 1963, habitat destruction, illegal hunting, and the effects of DDT had reduced the bald eagle population in the United States (not including Alaska) to just 147 breeding pairs.

## Anything but Silence

Carson spent 4 years gathering information and writing the book, *Silent Spring*, published in 1962. The title is based on her warning that we may be moving toward springs that are without bird songs if we do nothing to stop polluting the environment with DDT and other pesticides. The book quickly gained a lot of public attention and debate.

Chemical manufacturers were furious with Carson. They ran ads telling Americans to ignore her work and tried to discredit her ability as a scientist. They assured the public that pesticides were perfectly safe.

But Americans did worry. The evidence was quite convincing, and the White House and Congress were flooded with letters from anxious citizens demanding that something be done. President John F. Kennedy (1917–1963) called for a special committee of scientists to investigate Carson's claims. Congress also formed an investigative committee.

Carson spent the remaining years of her life defending the legitimacy of her findings. When she testified before the US Senate, she declared, "I deeply believe that we in this generation must come to terms with nature." In defending her research, Carson told Americans to think for themselves. "As you listen to the present controversy about pesticides," said Carson, "I recommend that you ask yourself: Who speaks? And why?"

## Responding to New Awareness

Carson's ideas may not seem revolutionary today. But before 1962, few people were familiar with terms like *pollution* or *environmental awareness*. US industries were constantly coming out with useful and exciting new products, and few people stopped to think if there could be negative side effects.

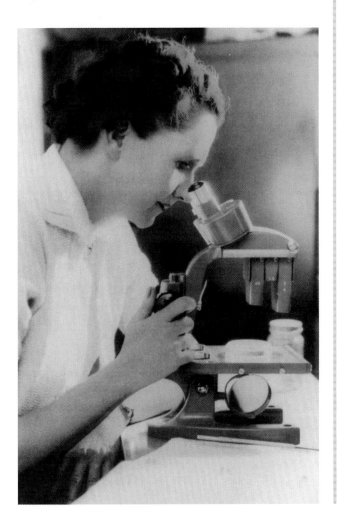

President Kennedy's committee found evidence that supported Carson's warnings. So did other government studies. Congress began passing laws to ban or control the use of potentially dangerous pesticides. In 1970, Congress established the Environmental Protection Agency (EPA) to reduce and control pollution of water, air, and soil. In 1973, DDT was banned in the United States.

The ban on DDT, combined with captive breeding and habitat protection programs, allowed bald eagle populations to recover. Today, there are an estimated 4,000 breeding pairs south of Canada.

After her death, Carson was awarded the Presidential Medal of Freedom by President Jimmy Carter (1924– ). President Carter described Carson as "a biologist with a gentle, clear voice, she welcomed her audiences to her love of the sea, while with an equally clear determined voice she warned Americans of the dangers human beings themselves pose for their own environment."

Can we avoid the silent spring that Carson predicted? In the more than 50 years since *Silent Spring* was published, people have become more aware of our impact on the environment. But we continue to invent new chemicals with unknown long-term impacts.

 **What is an environmental concern you have heard about in the news? Record the topic in your notebook, along with any details you think you know. What questions do you still have about this topic?**

# Trophic Levels

Organisms in an ecosystem can be grouped by their roles in the ecosystem. The roles are called **trophic levels**. Trophic levels are feeding levels. The trophic level at the base of the ecosystem is the producers.

The size of a trophic level can be measured in biomass. Biomass is the mass of matter produced by organisms in the ecosystem. Biomass includes organisms that used to be alive but are now dead. The producer trophic level always has the largest amount of biomass. The next largest amount of biomass is in the trophic level made up of consumers that eat producers,

the **primary consumers**. The primary consumers are eaten by the **secondary consumers**, and there are few of these. The secondary consumers are eaten by **tertiary consumers**, and there are even fewer of these. Each tropic level has less biomass than the level below it.

Look in your notebook to one example of a food web that you made in class. How could you organize the food web differently to show the relative number of producers and consumers?

# Food Webs

In a typical ecosystem, the producers, consumers, and **decomposers** (organisms that consume the remains of dead organisms and other biomass) transfer food energy from organism to organism in a model known as a food web. A food web, like those you have constructed in class, is a model that represents the flow of energy and matter through an ecosystem by showing all the feeding relationships. It also points out another important fact: some organisms must die to maintain the lives of others.

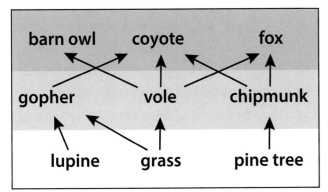

A food web

A food web uses arrows to connect organisms that eat other organisms. The arrow points from an organism to the organism that eats it. The arrows point in the direction that the energy in the food goes. When a fox eats a chipmunk, the food energy in the chipmunk goes into the fox. The arrow points from the chipmunk to the fox.

If the organisms are displayed by trophic level before arrows are drawn, the picture of community interactions is easier to see. In the example above, lupines, grass, and pine trees are producers. Gophers, voles, and chipmunks are primary consumers. They are also known as **herbivores**, because they eat only plants. Barn owls, coyotes, and foxes are secondary

consumers in their ecosystem. Secondary and tertiary consumers are also known as **carnivores**, or meat eaters, because they eat other animals. Some animals, including most humans, have a diet of both producers and consumers and are called **omnivores**.

Lupines are producers.

Chipmunks are primary consumers.

## Energy Transfer

What happens to the food made or eaten by an organism? Much of the energy available in food is used to maintain life functions. A portion of the energy is stored in the body of organisms in the form of biomass, but most is used to do work and then passes to the environment in the form of heat. So the energy needed to run, think, digest, pump blood, and perform all other life activities passes through the organism and into the environment.

The energy that passes through the ecosystem (entering as light from the Sun, transferring through one or more organisms, and exiting to the environment as heat) is often referred to as "lost" because it is not stored in biomass to be used as food at the next trophic level. Only a small percentage (sometimes as low as 10 percent) of the food consumed at a trophic level is converted into biomass at that level. As a result, consumers must eat a lot just to maintain their body mass and functions.

What does the difference between biomass eaten and biomass produced at a trophic level tell us about the number of organisms in each trophic level of an ecosystem? Will there be more producers? More primary consumers? More secondary consumers? This tells us that the biomass of producers will be much larger than the biomass of primary consumers. It also means that the biomass of primary consumers will be much larger than the biomass of secondary consumers.

## Trophic Pyramids

At each trophic level, the amount of energy transferred to new biomass through growth and reproduction is usually much less than the energy available in what the organism eats. In other words, there are usually fewer **predators** than **prey**. In the same way, there are generally fewer herbivores than plants. This is why trophic-level diagrams are often pyramids. Each layer is smaller as you go up the trophic levels, so the ecosystem diagram tapers to a point on top.

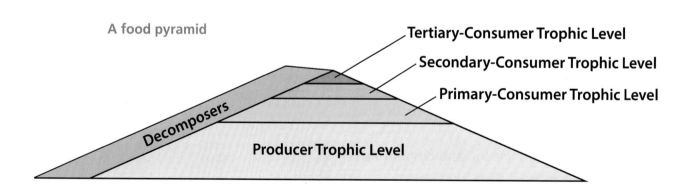

A food pyramid

Tertiary-Consumer Trophic Level

Secondary-Consumer Trophic Level

Primary-Consumer Trophic Level

Decomposers

Producer Trophic Level

Zooplankton

Anchovies

Algae

Orca

Salmon

Sea lion

When the ecosystem roles are placed in order from producers to primary consumer to secondary consumers, and so on, the result is a layered representation of the ecosystem. Each level of the **trophic pyramid** has less biomass than the one below it.

It's difficult to place the decomposers in the pyramid model of trophic levels because they affect every level. They are often placed along the side to suggest that they interact with all the other trophic levels.

It is extremely rare for an ecosystem to have fourth-level consumers that eat third-level (tertiary) consumers, because the amount of energy lost in each transfer is so great. Most examples of fourth-level (and higher) consumers live in freshwater aquatic or marine ecosystems. An orca is an example of a very high-level consumer. An orca might eat a sea lion, which ate a salmon, which

ate an anchovy, which ate zooplankton, which ate a single-celled alga. The orca is a fifth-level consumer. And when the great white shark eats the orca . . . you get the idea.

If you were organizing the organisms in an ecosystem into trophic levels, what would you do with the raccoons and crayfish? They both eat plant material, so they go in the primary-consumer level. But they both eat insects and other animals, too. That would place them in the secondary-consumer or possibly the tertiary-consumer level. Where do you place them?

Animals that are generalists, that is, those that feed at several trophic levels, should probably be represented in each trophic level where they play a role. Thus, the raccoons and crayfish might appear in three levels because they eat plants, insects, and fish.

## Pulling It All Together

Living organisms are complex. Their bodies are made of matter. The functions of living organisms are driven by energy. The energy for life comes from the Sun, captured in energy-rich molecules. The energy-rich molecules when consumed or used are called food.

Every organism needs a constant supply of matter and energy. Autotrophs (producers) get matter and energy from raw materials in the environment. Heterotrophs (consumers) get matter and energy by eating other organisms. Matter and energy move up through the trophic levels in an ecosystem by feeding relationships.

Dead organic material still has valuable matter and energy. It is the decomposers that get the last bit of energy out of organic material and reduce the molecules of matter to simple chemicals. The matter that entered the ecosystem as food made by producers returns to the environment to be used by living organisms again. Matter recycles again and again and again. But the fate of virtually all energy that passes through the trophic system is to be radiated into the environment as heat. Energy passes through the ecosystem only once. Once released to the environment, it is gone.

### Think Questions

1. What does the following sentence mean? "Matter recycles again and again, but energy passes through the ecosystem only once."

2. One model of an ecosystem is a trophic pyramid. Explain why the bottom layer of the pyramid is biggest, and the top layer is the smallest.

3. Why doesn't 100 percent of the energy pass from trophic level to trophic level?

# Decomposers

Many producers and consumers die without being eaten by other organisms. What happens to their bodies? Deciduous trees lose their leaves every year. Where do these leaves go? Animals produce feces. What happens to their waste? Dead organisms, dead parts of organisms, and feces are made of matter. This biomass collects on the ground (or the bottom of a lake or the ocean) and is available for other organisms to eat. Bacteria and fungi consume this organic matter. These specialized consumers are known as decomposers.

When something rots or decays, it is actually being consumed by decomposers. This process of decay transfers every last bit of energy from the dead organism to the decomposers. When the decomposers are through, all that remains are a few simple chemicals, which become soil, water, and carbon dioxide in the air. These simple chemicals are matter, which can be used once again by producers to make food during photosynthesis.

An oak tree without leaves

Bacteria and fungi are the most important decomposer organisms, and their role in the ecosystem clearly fits the description of decomposer. But what about organisms like worms, maggots, termites, and even vultures and coyotes, which consume dead organic matter for the energy they need? Are they decomposers, or are they herbivores and carnivores? Ecologists are still defining the roles played by the less glamorous, but critically important, organisms that clean up the dead and discarded bits of life.

A thriving community of organisms works on the film of dead organic matter that covers much of the surface of Earth at the interface between land and air or land and water. This organic matter, known as **detritus**, is the home of the **detritivores**, organisms that eat dead material. Not all detritivores reduce the matter to simple chemicals. Worms and beetle larvae, for instance, miss a significant amount of matter and produce feces, which in turn are consumed by other decomposers who can use the energy left in the feces.

Decomposers provide many ecosystem services. In addition to recycling dead matter, they make important nutrients available again to organisms. Without decomposers, the ecosystem's recycling crew, Earth would be buried under mountains of waste and dead organisms!

A vulture

Termites

Maggots may seem disgusting, but they perform an essential role in helping make nutrients available in an ecosystem for organisms to use.

# Limiting Factors

Nothing lives forever. Even the ancient bristlecone pine trees in the high mountains of the western United States die after a few thousand years. Most organisms live much shorter lives. Many insects live a few months; fish and small mammals a few years; many plants, reptiles, birds, and large mammals a few decades; and a scattering of others, like trees, a few centuries. Life is a temporary thing . . . for the individual.

If a species is to continue to exist on Earth, the species must produce new individuals continually. Producing new individuals to maintain a population is reproduction, and every species has a way of reproducing.

The rate at which a species can increase its population is its **reproductive potential**. Some species, like elephants, have modest reproductive potential. A female elephant can produce a single offspring every 4 years. A single female Atlantic cod, on the other hand, can lay 10 million eggs in 1 year. Clearly the potential for the cod to increase the cod population is much greater than the potential for the elephant to increase the elephant population.

**Atlantic cod**

So why don't populations spiral out of control? Why aren't there billions of billions of trillions of Atlantic cod filling the ocean from top to bottom after 5 or 10 years? Because there are **limiting factors** for every population on Earth. Abiotic and biotic limiting factors control the size of populations.

## Biotic Limits: Predation

When one organism is eaten by another, that is a limiting factor. Every organism is desirable to some other organism as a source of food. As you know, food provides the energy that is essential for survival. When an individual reproduces, its offspring are new biomass. Predators may take advantage of this new energy source. We see this kind of population control in Mono Lake when the brine shrimp feed on the planktonic algae, reducing their numbers, and in turn the phalaropes and gulls eat brine shrimp, reducing their population. Predation can occur at any stage in the life cycle of an organism. When predators eat individuals, the population size is limited.

Humans, while being omnivores, can also be considered predators. We collect fish by the boatload and can harvest a field of wheat in a day. Sometimes we hunt for sport and don't eat the animals we kill. Another notorious predator is free-ranging house cats. They are responsible for killing about 1 to 4 billion birds and 6 to 22 billion mammals in the United States every year.

Harvesting a wheat field

## Biotic Limits: Disease

Diseases limit populations in the same way. Even though we don't usually think of a large animal or plant being attacked by a microscopic bacterium, the result can be the same. A mountain lion capturing a deer removes an individual organism from the population. A disease organism can enter a population and kill large numbers of individuals.

Now that humans can travel from the depths of a jungle to a populated city across the world within 1 day, infectious diseases can spread more easily. Diseases are becoming a more significant limiting factor for human populations and for many other organisms with which we interact.

A mountain lion

66

## Biotic Limits: Resources

Populations are limited by food supply. If an organism cannot acquire the energy needed to survive and reproduce, the population will decline and, with it, the potential for producing the next generation. If a snake cannot find enough mice to sustain itself, it will starve to death. Even if it survives, it may be so weak that it can't reproduce. Similarly, if there is a poor crop of acorns, squirrels may starve. Even if they survive, they may not be able to feed their young. In 1982, a reduced population of brine shrimp in Mono Lake prevented the California gulls from successfully feeding their chicks. Most of the gull offspring died that year. Lack of food is one of the most important limitations on populations.

## Abiotic Limits: Reproductive Environments

Many organisms require specific conditions in order to reproduce. If the number and quality of locations where reproduction can occur are limited, reproduction will be limited. Bank swallows need sandy cliffs in which to dig nesting burrows. If a sandy cliff tumbles down during a flood or earthquake, suitable nesting sites are lost. Salmon need clean gravel streambeds in which to lay their eggs, and black bears need winter dens in which to give birth. Without an environment that provides for the physical conditions needed to reproduce, young will not be born.

A bank swallow

Bank swallows made nesting burrows in this sandy cliff.

An American bison

Another way to consider the reproductive environment is habitat availability. As humans clear forests and grasslands and coastal ecosystems to expand cities and farmlands, we change the habitats where organisms live. Sometimes the habitats are no longer suitable for certain populations to reproduce. Habitat destruction is a serious limiting factor for many organisms, particularly many currently listed as in danger of **extinction**.

## Abiotic Limits: Seasons

Seasonal changes put pressure on populations. In the temperate and polar latitudes, winter is a major factor in population limitation. During winter, days are shorter, so production by photosynthetic organisms slows or, for some trees, stops entirely. Often, winter brings rain, snow, and wind, each of which puts stress on populations.

Some animals respond to the threat of wind, flood, and freezing by leaving the area. Birds, because of their mobility, are famous for migrating to warmer regions. Others, like the American bison and caribou, go on long or short treks to find suitable winter environments. Some organisms become dormant, basically shutting down until spring. Frogs, fish, bears, squirrels, snakes, maple trees, and many other organisms use dormancy, reduced activity, and winter sleep to wait out the winter. These strategies work if the wintering place offers sufficient protection and the organism has accumulated enough fat or has stored enough food to survive the winter.

Winter is the main limiting factor for many temperate and polar populations. Many populations decline to minimal levels, like the brine shrimp in Mono Lake, and then expand rapidly in the spring. Seasonal changes in population size such as those at Mono Lake are normal and healthy.

## Abiotic Limits: Climate Change

Climate change is an important abiotic limiting factor for many species. Earth's climate changes over time, but it has been changing more quickly over the last 100 years as the concentration of greenhouse gases in the atmosphere has increased. These gases (including carbon dioxide, $CO_2$) effectively trap heat within Earth's atmosphere, making the average global temperature higher. Climate change is predicted to cause increases in species extinction over the next 100 years.

Mountain and arctic ecosystems and species are particularly sensitive to climate change. One consequence of climate change is a rise in sea level. As it rises, salt water can seep into freshwater systems and force organisms to relocate or die. It also means less suitable habitat for land animals.

Climate change can alter where species live and how they interact, which could transform current ecosystems. As you know, impacts on one species can ripple through the food web and affect many organisms in an ecosystem. For many species, the climate influences timing for migration, blooming, and mating. As climates change, the timing of these events has also changed in some parts of the world. This can lead to mismatches for other organisms, limiting their populations. Growth and survival are reduced when animals arrive at a location before or after their normal seasonal food sources are present.

## Carrying Capacity

When you stand back and take in the large view of life on Earth, you realize it is a struggle to survive here. Every living thing has fundamental requirements for life, and if it doesn't get those things, it dies. One of the most critical requirements is energy.

Energy enters the ecosystem as sunlight, or solar energy. Photosynthetic organisms capture the energy and transform it into carbohydrates, like sugar, that we call food. The energy is stored in the bonds of the food molecule. The amount of food that can be produced is limited by several factors, including light, space for living, and availability of water, carbon dioxide, and

Sunlight provides energy for the ecosystem.

minerals. For any given ecosystem, there is a limit to the amount of food that the producers can make.

We know that the other organisms in an ecosystem acquire energy by eating each other. Primary consumers eat producers, secondary consumers eat primary consumers, and so on. The number of consumers is limited by the amount of production by producers.

The total number of individuals that an ecosystem can sustain indefinitely is the **carrying capacity**. For instance, a backyard ecosystem might support three rabbits year after year on the amount of grass and other vegetation growing there. The carrying capacity for rabbits seems to be three. If six rabbits move in, the carrying capacity of the ecosystem might be exceeded. As a consequence, in order to survive, the rabbits will eat so much of the vegetation that they could wipe out the producer populations. Exceeding the carrying capacity of an ecosystem produces changes that will alter the nature of the ecosystem.

In the rabbits example, if they eat so many producers that the producers can't reproduce, the food source for the rabbits would be depleted. The consumers' offspring (baby rabbits) would have nothing to eat, so the consumer population would die off. In **sustainable** ecosystems, there must be survivors of every species in sufficient numbers to reproduce and keep the population going.

Rabbits

Usually the amount of primary production establishes the overall carrying capacity of an ecosystem. Primary production is the amount of food (made by the photosynthesizers) that can be consumed and redistributed throughout the food web of the ecosystem over the course of a year. When you know the amount of primary production, you are close to knowing the carrying capacity of an ecosystem.

**Review your observations of the classroom habitats. What are some limiting factors in each habitat?**

Mono Lake

Another factor is the time needed for the producers to make food. Consider the carrying capacity of Mono Lake. The lake has plenty of light, water, carbon dioxide, and minerals. The producers of the lake, the algae, reproduce rapidly. Food production in the lake is extremely high. The amount of algae produced in the lake supports trillions of brine shrimp and brine flies. These in turn nourish millions of birds and a few coyotes. A lot of life flows through Mono Lake each year.

Contrast this with the time needed to produce food on the Great Plains, another very productive ecosystem. Grasses grow more slowly, taking longer to produce food. The amount of grazing by insects, rodents, deer, and cattle must not exceed the capacity of the grasses to regenerate. Each ecosystem has its own carrying capacity that depends on the ability of the producers to support the ecosystem food web.

A grassland environment

## Think Questions

1. What are some biotic limiting factors?

2. What are some abiotic limiting factors?

3. What are some limiting factors for your ecoscenario? Hint: Think about what you've learned about your ecoscenario so far, and consider temperature, food sources, and other factors mentioned in this article.

# Mono Lake throughout the Year

**Winter.** During the cold of winter, Mono Lake looks like a wasteland. No birds swim on its surface, and no brine flies or brine shrimp are active. However, the lake is turning a healthy dull green color, as the planktonic algae experience an explosive phase of population growth.

The algae population increases rapidly because for a while there are no limits on its growth. Light, nutrients, and carbon dioxide are plentiful. The algae are not limited by the near-freezing temperature of the water, and most of the predators are dormant.

**Spring.** As spring arrives, the increased solar energy warms the water. Brine shrimp and brine flies hatch and start eating the algae. The hungry predators eat the algae faster than the algae can reproduce, so the algae population starts to decline.

**Summer.** As summer approaches, the first birds, the gulls, arrive. They feast on the brine shrimp. The gulls slow the population growth of the brine shrimp, but don't reverse it. It is not until July, when all the migratory birds are present, that the brine shrimp begin to decline. They decline because they are reaching the limit of their food supply on one side, and hundreds of thousands of birds are eating them on the other side. The brine shrimp population declines rapidly.

California gulls migrate to Mono Lake to feast on flies and shrimp.

**Fall.** By October, the nutrients in the water have been depleted, and predation has taken its toll on the algae population. It falls to its lowest level. The cooling water causes the eggs laid by the remaining brine shrimp and the pupae of the brine flies to fall into dormancy, and the adult brine shrimp and brine fly populations drop to zero.

When the surface water approaches freezing, it is so dense that it plunges to the bottom of the lake, stirring up the nutrient-rich sediments, making them available to the algae. The algae begin, once again, to reproduce rapidly, turning the lake green. The producers have set the stage for a repeat of the age-old surge of life in the Mono Lake ecosystem.

Frozen Mono Lake

# Biodiversity

**Biodiversity** is the variety of life. Biodiversity refers to all the different kinds of organisms living in a given area. It can be studied at many levels. It could include the entire Earth and all living organisms on Earth, or just a portion of the schoolyard.

Let's look at the biodiversity of a pond. At first glance, we notice the animals: turtles sunning themselves; dragonflies zipping by; different kinds of fish swimming; and a red-winged blackbird singing. Getting closer, we notice smaller creatures like leeches and water striders. The algae, cattails, and water lilies in the pond, and the tall grasses, shrubs, and trees surrounding the pond contribute to the biodiversity of this ecosystem.

You need a microscope to complete your inventory of the biodiversity of the pond. Using a microscope, you would see hundreds of thousands of different kinds of microorganisms suspended in the pond water.

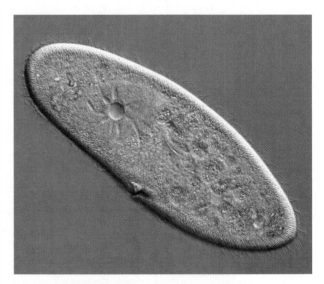

**Microorganisms inhabit every drop of pond water.**

**Aquatic biodiversity**

A common way to measure biodiversity is to count the total number of species living within a particular area in a process called **sampling**. A **biodiversity index** can then be calculated as a measure of the health of an ecosystem. Scientists have identified around 1.5 million species, but estimate that there are 5 to 100 million species on Earth! Most of the species yet to be identified are microorganisms.

Typically, regions near the equator have the greatest biodiversity. This is true for both terrestrial and aquatic biomes. Tropical biomes are warm and sunny with abundant rainfall year-round. This provides energy and water in a stable environment for a huge number of species. Biomes that have severe weather (hot and cold) or limited precipitation have less biodiversity. Fewer species can live in these environments.

## Importance of Biodiversity

All species in an ecosystem are **interdependent**. If the biodiversity is low and a population dies off, the entire ecosystem may fail. When the biodiversity is high, there is a better chance that the ecosystem can recover if something disturbs a population in the food web.

Ecosystems with high biodiversity are able to recover faster from natural disasters (extreme storms, fires, or earthquakes) than the same kind of ecosystem with lower biodiversity. Because high biodiversity helps an ecosystem recover from disruption, high biodiversity helps humans continue to benefit from ecosystem services such as food and clean water.

Not every species contributes equally to biodiversity. A **keystone species** is critical for the overall health of an ecosystem. The loss of that species creates a chain of events that can impact the biodiversity of the entire ecosystem.

When keystone species are top predators, they keep populations in a food chain balanced. The wolves in Yellowstone National Park and sea otters in Monterey Bay National Marine Sanctuary are keystone species. Without these predators, their prey populations can grow too large, eat too much, and kill off other populations in the ecosystem.

Bees collect pollen from flowers.

Bees and insects carrying pollen from one plant to another are also keystone species. These pollinators are important to maintaining the health of the ecosystems they live in. Bees, birds, and other creatures pollinate 75 percent of the world's food crops.

**Ecosystem engineers** are a special kind of keystone species. They change the ecosystem they live in. Beavers are an example of an ecosystem engineer. By building dams in rivers and streams, beavers create meadows where many kinds of plants can grow and animals can live. Earthworms that churn topsoil and whales that churn ocean waters when feeding are also examples of ecosystem engineers.

## Threats to Biodiversity

The major threat to biodiversity is extinction. A species becomes **extinct** when the last of its members no longer reproduce and the population disappears. All the other organisms connected to it in an ecosystem are affected. If those organisms relied on the species for food, they must find an alternative food source or die out themselves. The loss of a keystone species is particularly devastating to an ecosystem.

Extinction is a part of life. Paleontologists estimate that 99 percent of species that have ever existed on Earth are now extinct. There have been five major extinctions in Earth's history, caused by meteor impacts, volcanic eruptions, and natural climate change. Scientists believe we are living in a period of the sixth major extinction, one caused by humans. Humans have significantly affected the size of populations on Earth for at least 10,000 years. With human population growth, many of these changes have become more rapid in the past few centuries. Scientists estimate that one-third of all known living species are currently threatened with extinction. Here are some key ways in which humans impact biodiversity.

**Habitat loss.** A major factor that reduces biodiversity is change or destruction of habitats. Organisms may not be able to survive outside their original habitat, so biodiversity decreases. Humans change habitats mainly through

- construction,
- agriculture,
- deforestation (cutting down forest trees),
- dragging fishing nets in the ocean and bays,
- damming or rerouting streams and rivers, and
- draining and paving over wetlands and estuaries.

Hoover Dam

Deforestation

**Overhunting/overfishing.** Humans are predators. We eat other animals for food. Sometimes too many animals in a population are killed. This is called overhunting or overfishing. When this happens, a population of food animals may die out of a particular ecosystem or even go extinct, reducing the ecosystem's biodiversity.

**Invasive species.** When humans introduce organisms into an ecosystem they can disrupt the ecosystem. These **invasive species** can decrease biodiversity by competing for resources with organisms that lived there before.

**Pollution.** Chemicals or other substances that are introduced into an ecosystem can harm organisms. Pollution can decrease the biodiversity of an ecosystem by damaging or killing organisms.

Air pollution

**Climate change.** Changes in abiotic factors (like temperature and precipitation patterns) of a place can have dramatic impacts on the survival of local plants and animals. In addition, rising sea levels due to climate change can reduce land space available for plants and animals, and can add salt to freshwater biomes. If populations can't survive the changes, they die off, decreasing biodiversity.

# Protecting Biodiversity

Learning that humans are driving species into extinction is a heavy burden. But here is the good news: we are also the only species that can use science and technology to change our impact. We have an opportunity to have a positive impact on biodiversity. Here are some ways that you can think about your own impact on biodiversity.

**Think about what you eat.** The best place to start is with what you personally consume. Eat lower on the food chain. What does that mean? Eating meat can affect an ecosystem more than eating plants. It takes a huge amount of energy and resources to raise animals for food. This doesn't mean you need to become a vegetarian, but be aware of what you eat. You can be more of an omnivore than a carnivore.

**Reduce, reuse, recycle.** We consume more than just food. We also consume a lot of natural resources such as metal and wood. Obtaining more of these materials affects ecosystems. Instead, consider recycling materials like cans and paper, and buying products made from recycled materials. Reusable grocery bags, reusable water bottles, and coffee mugs reduce the consumption of new materials.

**Think global, act local.** It is easy to get overwhelmed when you hear about big environmental problems like deforestation and climate change. Remember that every world leader started by seeking local solutions. Consider small ways to improve your backyard, schoolyard, and neighborhood. Volunteer with a local organization or start your own. "Roots and Shoots," founded by Jane Goodall (1934– ), is a great online resource to help you get started with local environmental change.
*For more information on the Roots and Shoots program visit FOSSweb.*

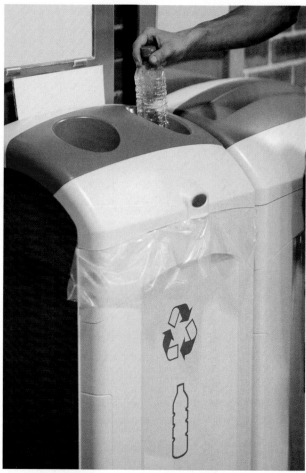

Recycling is a great first step in conserving our resources.

European starling

# Invasive Species

A **native species** is a kind of organism that has been part of a community for a long time. Species that are brought to a place by humans and become part of the local ecosystem have several names. We refer to them as **introduced species**, but they can also be called nonnative or exotic species.

Introduced species can change the balance of an ecosystem. Not all organisms introduced to a new ecosystem survive. And not all introduced species that survive will have negative effects on native species. But when they do, it can be disastrous. When an introduced species has a negative impact on an ecosystem, it is labeled an invasive species.

How are organisms introduced to new ecosystems? Humans sometimes introduce new species to an ecosystem on purpose. In 1890, 60 European starlings were released into New York's Central Park as part of a Shakespeare festival. Without natural predators, the rowdy birds survived and flourished. Today, millions of starlings live in North America, and often cause problems by outcompeting native birds for nest spaces.

What are some invasive species you have heard of? Record a list in your notebook.

Most invasive species are accidentally introduced to a new ecosystem. Many aquatic invasive species hitched a ride to a new location while attached to the bottom of ships or in a ship's ballast tanks (water tanks used to stabilize the ship). Global trading of natural goods, such as timber, plants, and pets, has allowed many species hidden within the leaves or fur of the organism to go along for the ride.

## Impacts of Invasive Species

An invasive species can change habitats and alter ecosystem function and ecosystem services. In fact, invasive species are a greater threat to native biodiversity than overhunting, pollution, and disease combined. Interactions with invasive species are the main threat for half of the species in the United States that are at risk of extinction.

Invasive species threaten biodiversity and ecosystem services by
- crowding out or replacing native species,
- interfering with human activities, such as raising livestock or crops,
- carrying and transmitting disease,
- acting as predators or parasites, or
- changing the existing habitat.

## Major US Invasive Species

**Powerful predators.** New predators introduced to an ecosystem can dramatically affect the balance of the existing food chain. If they are too successful, they can reduce other populations, perhaps even driving them extinct. That will, in turn, cause other effects throughout the ecosystem.

The *black rat* is native to tropical Asia. Rats hitched rides on ships as long ago as 100 CE and thrived in the growing cities where ships would dock. These predators will eat nearly anything, but particularly enjoy bird and reptile eggs and baby animals. Because of this, rats are believed to have caused the extinction of many bird, reptile, and other small animal species. In addition, fleas living on rats are able to carry infectious diseases that can be deadly to humans. The plague, transmitted by fleas on rats, killed 75 to 200 million people in the 1300s.

A black rat

Free-ranging *domestic cats* are one of the greatest threats to small mammal and bird biodiversity across the world. In the United States alone, domestic cats kill up to 4 billion birds and 20 billion mammals every year. Recent studies suggest that domestic cats kill more native birds and mammals in the United States than any other human impact (including collision with buildings and cars, and hunting by other invasive predators introduced by humans).

A domestic cat

The *northern snakehead fish* seems like the perfect invasive predator. Native to the waters of Southeast Asia, this fish will eat anything in its way. It took only a few individuals released into the eastern US watersheds to cause trouble. What makes them so dangerous to freshwater biodiversity is that they can survive outside of water for several days. Young fish can travel by wriggling on land to search for a new home! In addition, each mature female can lay up to 75,000 eggs each year.

The *brown tree snake* was introduced accidentally to the island of Guam through cargo shipments. (Interestingly, a 2.5 meter (m) long brown tree snake survived for almost a year packed in an unsuspecting homeowner's washing machine.) Within about 20 years, the introduced snake population eliminated 10 out of 11 bird species native to the forests of Guam. Today, you cannot find a native bird on the island.

*Lionfish* came to the United States from the western Pacific and the Indian Ocean as aquarium fish. As they grew too large, some owners released them into the Florida coastal waters. With no natural predator and plenty of fish to eat, the lionfish population quickly

A lionfish

grew. Some areas offer contests in an attempt to encourage lionfish fishing, and marine sanctuaries have asked snorkelers to report lionfish sightings in order to help managers prevent further population growth.

**Competitors and harassers.** Some introduced organisms outcompete native organisms for food or habitat space within the ecosystem. *Zebra mussels* were accidentally brought to the United States from Russia in the ballasts of ships. Without any natural predators, zebra mussel populations have grown dramatically. The mussels cluster on any underwater surface, clogging pipe systems and costing billions of dollars in damages each year. At least 30 species of native freshwater mussels are threatened with extinction because of competition with the zebra mussel. The ecosystems of the Great Lakes have been significantly changed by zebra mussels.

A brown tree snake

A zebra mussel

The North American *gray squirrel* is a native species of eastern United States deciduous forests. It was transported to Europe by humans, where it made a new home in the European deciduous forests. This squirrel is larger and naturally more aggressive than the native European red squirrels. It's driving them to extinction by outcompeting them for the same food sources.

*Kudzu* is a climbing vine native to southern Japan and southeast China. Its growth is out of control in the southeastern United States. Also known as the mile-a-minute vine, this fast-growing plant was intentionally introduced to help inhibit soil erosion. Without any natural predators, kudzu has literally covered many ecosystems.

A gray squirrel

Kudzu has grown over an entire area.

**Microscopic invaders.** Organisms that aren't easily visible can be just as dangerous when introduced to a new ecosystem. *Chestnut blight fungus* came to the United States in the early 1900s on trees imported from Asia. The fungus attacked American chestnut trees, wiping out nearly 4 billion trees within 50 years. Less than 100 mature trees are estimated to be alive in the tree's main eastern United States ecosystem. The near-extinction of the American chestnut was a disaster for the many animals that depended on the tree for habitat or food.

## Minimizing Impacts on Biodiversity

The total economic impact of invasive species in the United States is hard to calculate, but costs many billions of dollars in terms of lost crops or products and efforts to combat the invasive species. We use three main ways to minimize the impacts of invasive species.

**Prevention.** Keeping a known invasive species out is the best strategy to protect an ecosystem. Because humans are the main transporters of invasive species, we can reduce this part of the problem with education and regulations. Many boat licenses require owners to clean boats before launching them in new bodies of water. Border customs regulate the import of invasive species and materials that could carry invisible hitchhikers.

**Eradication.** Once an invasive species has arrived in an ecosystem, the next step is to try to eliminate it before it becomes uncontrollable. Early efforts can help reduce the overall impact on the ecosystem.

An Indian mongoose

Sometimes humans introduce other organisms that eat or compete with the invasive population. However, introducing another organism can cause even more troubles. One example is the introduction of the Indian mongoose into New Zealand to control invasive rat and opossum populations. Unfortunately, the mongoose preferred hunting the native birds. Today, New Zealand has three invasive predators instead of two.

**Mitigation.** Once an invasive species has settled into an ecosystem, humans may need to spend a lot of money to protect the native species, crops, and livestock.

Invasive species can harm crops in many ways.

## What Can You Do?

You can help support native species in your area.

**Plant a native garden.** Many invasive plants were brought to the United States to make our gardens look like gardens in other places around the world. We can plant instead the many beautiful native plants that provide food and shelter for native animals. Because native plants are well adapted to their ecosystems, they also have no need for excessive watering or pesticides.

**Practice prevention strategies.** Avoid transporting plants, fruit, or animals to new places, especially when signs warn against it.

**Keep your cat indoors.** For most Americans, the most important way they can help protect native biodiversity is to keep the family cat indoors. Studies have found that putting a bell on a cat's collar is not helpful, as most wildlife will not identify the sounds with danger until it is too late. You can also help by supporting your local animal shelter.

Many shelters help find homes for feral cats, reducing their hunting activities. They also curb cat populations by spaying female cats.

**Support habitat restoration projects.** You can get involved in projects to help restore natural areas. These projects may include working outdoors to remove invasive species, to plant more native species, or to remove pollutants. By minimizing human impact on an area, we can give native plants and animals a better chance of surviving.

Dahlias are native to Mexico.

Is it a good idea to introduce your classroom organisms into the local ecosystem at the end of the year? Why or why not?

# Mono Lake in the Spotlight

There is really only one issue at Mono Lake . . . water. Life depends on water in Mono Lake as surely as it does in every other ecosystem on Earth. Humans have a tremendous thirst for water, and whenever they see an opportunity to gather some up for human use, they usually do.

## Water Use at Mono Lake

The Los Angeles Department of Water and Power (LADWP) looked at the streams running out of the Sierra Nevada mountains into what seemed to be a useless, salty lake— Mono Lake. They decided to build a system of dams to divert the water to Los Angeles before it went into the lake. The project was completed in 1941, and virtually all the water that had flowed into Mono Lake was piped to Southern California. Without the annual inflow of fresh water, the lake began to dry up. The surface of the lake was at 1,956.5 meters (m) elevation in 1941. By the 1960s, the lake surface had dropped by 6.5 m.

California gulls feeding at Mono Lake

By the mid 1970s, the lake level was down more than 12 m. The lake held just half the amount of water it had before its water sources were rerouted. Changes started to appear in the ecosystem. The salt was twice as concentrated, and this limited the populations of primary consumers, the flies and brine shrimp. The shrimp were not growing as large as usual, and their numbers were declining. In 1982, the brine shrimp production was so reduced that the 50,000 breeding California gulls could not catch enough shrimp to feed their offspring, and 25,000 half-grown chicks starved to death. Furthermore, the water was so low that a land bridge developed between Negit Island and the mainland. The bridge allowed predators like coyotes to walk to the nesting area, where they ate the gull eggs and chicks and drove the adults away.

David Gaines

## A Plan for Restoration

In 1978, David Gaines (1947–1988) became concerned about the rapidly changing conditions in the Mono Lake ecosystem. He founded an action group called the Mono Lake Committee. He started to work on ways to reverse the damage done to the Mono Lake ecosystem. He worked tirelessly with the

National Audubon Society, California Trout, the California Department of Fish and Game, the US Forest Service, and the LADWP to find solutions to the problem.

In 1994, after years of negotiations, a landmark California Supreme Court decision ruled that the LADWP should release enough water to restore the lake to a level of 1,948 m over about 20 years, and then to keep the lake at that level. This was not as high as the lake used to be, but it was a compromise between the needs of the ecosystem and the needs of people. With this compromise, there is some water for Los Angeles, the Mono Lake ecosystem is sustainable, there is no land bridge to Negit Island, trout can spawn in the freshwater

A coyote

creeks feeding Mono Lake, and there are excellent scenic views for Mono Lake visitors.

The lake today is very close to the goal of 1,948 m. The replenished freshwater sources have brought the salinity back to historical levels. The plan for water use remains a continued conversation between Mono Lake managers and the LADWP. In Los Angeles, much of the water lost to Mono Lake is not actually missed, because the city has improved water conservation, education efforts, and water recycling. In one successful city plan, the LADWP replaced old toilets throughout the city with new low-flush models. They saved more water in 1 year than was ever diverted from Mono Lake!

## Conserving Water in Your Ecosystem

Some places in the United States have more limited water resources than others. Water may not be a limited resource in your area, but using simple water-conservation habits is a responsible choice that you can make to help support your local ecosystem.

**Showers.** Showering is one of the main ways we use water in our homes. The average showerhead uses 10 liters (L) of water per minute. Installing a water-saving showerhead doesn't change the feel of the shower pressure but can save your family 11,600 L of water each year! You can save another 800 to 1,200 L of water a year if you collect the cold water as you wait for your shower to warm up and use this water in your garden.

### How Much Water Do We Use?

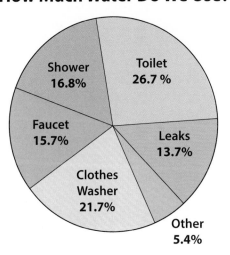

**Faucet.** The average faucet flows at 8 L per minute. If you could prevent 1 minute of water flow each day (turning off water while brushing your teeth, shaving, or scrubbing dishes), that would be over 2,800 L per year! The more strategic you are about turning the water off and on, the more water you can save.

**Dishwasher.** The average dishwasher uses just 16 to 24 L of water for a full washing cycle. Scrape any food off dishes with a spatula instead of rinsing, to save water.

**Drinking water.** Many people run the tap until it is cool for drinking. If you fill a bottle with tap water and store it in the refrigerator to keep it cool instead, you can save 800 to 1,200 L of water per year.

**Toilet.** Encourage your family to install low-flush toilets, which use only 6 L per flush.

**Gardening.** Watering the lawn in the morning instead of in the heat of the day can save water that would be lost to evaporation. Also avoid watering on windy days, which causes extra evaporation. A layer of mulch around trees and plants can reduce evaporation significantly. Consider installing a rain barrel, which can collect water that runs off your roof to water your garden later. Plants that are native to your area often need less maintenance and less water, and support local wildlife.

**Rain barrels**

# Images and Data

# Images and Data Table of Contents

# Minihabitat Organisms

## Earthworm

**Genus:** *Lumbricus*

**Species:** Varies

**Size:** Up to 25 cm long

**Range:** Worldwide except polar regions

**Natural History:** Earthworms live in the upper layers of the soil but will tunnel as deep as 2 m if conditions are too dry or too cool. They prefer light, loamy soils to those high in clay and sand. Temperatures of about 13°C are ideal.

**Food:** Earthworms eat detritus. They decompose the organic material and return nutrients to the ground.

**Predators:** Birds, frogs, toads, salamanders, lizards, shrews, minks, raccoons, and turtles

**Shelter:** Tunnels in the upper layers of soil

**Reproduction:** Earthworms are hermaphroditic (have both male and female sex organs). The mating pair fertilizes each other. The resulting cocoon containing eggs is left behind in the soil when the pair separates. Tiny earthworms emerge in 2–4 weeks.

**Abiotic Interaction:** The movement of earthworms helps break up, loosen, and churn topsoil.

**Biotic Interaction:** Humans add earthworms to gardens to improve the quality of the soil. Earthworm waste products enrich the soil and recycle nutrients into the environment. Earthworms are also used as fish bait.

## Aquatic snail

**Genus:** *Planorbis*

**Species:** Varies

**Size:** Shell up to 3 cm diameter

**Range:** Temperate and tropical freshwater ponds

**Natural History:** Snails have a hard, spiraled shell. It gets bigger toward the opening as the snail grows. The muscular part that sticks out from the shell is the foot. The snail scrapes algae from the surfaces over which it travels.

**Food:** Algae, aquatic plants, and detritus. Snails eat whatever food is left over by fish.

**Predators:** Large fish, birds, waterfowl, shrews, and turtles

**Shelter:** If threatened, an aquatic snail pulls into its shell and closes the opening with a door made of hard material, called an operculum.

**Reproduction:** Snails are hermaphroditic (have both male and female sex organs). During mating, two snails exchange sperm, and both may lay eggs. Eggs are laid on plants in a jelly capsule. The eggs hatch into tiny larvae that swim freely until they begin to grow a shell. The shell weighs them down, and they begin a life of crawling along the pond bottom.

**Abiotic Interaction:** When there is less calcium in the environment or an increase in acidity, snail shells can become weak. Changes in water quality from human activities (such as runoff of fertilizer into ponds or streams) can directly affect the snails' habitat.

**Biotic Interaction:** Aquatic snails are kept in home aquariums to reduce the amount of algae and detritus.

## Isopod

**Genus:** *Oniscus*

**Species:** Varies

**Size:** 1.8 cm

**Range:** Temperate and tropical regions

**Natural History:** Isopods are found in dark, damp areas, especially under rocks and in leaf litter.

**Food:** Detritus, fruit, fungi, and young plants

**Predators:** Birds, frogs, lizards, turtles, and salamanders

**Shelter:** Under rocks and logs, and in leaf litter. Some species roll into a tight ball if threatened.

**Reproduction:** Females carry eggs until they hatch.

**Abiotic Interaction:** Isopods must have a dark, moist habitat.

**Biotic Interaction:** Isopods can be a minor agricultural pest by eating plants, but they also enrich soil by recycling nutrients.

## Guppy

**Genus:** *Poecilia*

**Species:** *reticulata*

**Size:** 3 cm

**Range:** Freshwater streams of Central America and northern South America

**Natural History:** Guppies are small fish that bear live young. Females are usually beige or silver gray. Males are smaller and have longer, flowing tails.

**Food:** These omnivores eat detritus, algae, smaller fish, and plants.

**Predators:** Adult guppies eat unprotected young. In their natural environment, larger fish of other species are predators of guppies.

**Shelter:** Females and young hide in vegetation. Guppies can live in an unheated aquarium unless room temperatures go below 15°C.

**Reproduction:** Guppies breed easily in home aquariums. Females can store sperm and may produce several broods from one mating. Babies are born live and must immediately find shelter to avoid being eaten.

**Abiotic Interaction:** Guppies are freshwater fish, which means they cannot survive in a saltwater habitat.

**Biotic Interaction:** Brightly colored guppies are bred for home aquariums. Feeder guppies are raised as food for larger aquarium fish. Guppies released from aquariums can disrupt the predator/prey interactions in local aquatic environments.

## Scud

**Genus:** *Gammarus*

**Species:** Varies

**Size:** 5–30 mm

**Range:** Fresh water in the Northern Hemisphere

**Natural History:** Scuds are much more active at night than during the day. They crawl and walk using their legs, in addition to flexing their body.

**Food:** Bacteria, algae, and detritus

**Predators:** Fish, toads, salamanders, waterfowl, and other crustaceans

**Shelter:** Detritus or plant material. Scuds usually live close to the bottom or among submerged objects where they can hide from predators. They prefer dark areas.

**Reproduction:** Most scuds breed between February and October. Eggs and young develop in a brood pouch on the female. Young stay in the pouch about a week or so.

**Abiotic Interaction:** Scuds are a freshwater organism, which means they cannot survive in a saltwater habitat.

**Biotic Interaction:** Scuds are found in almost all aquatic environments and help recycle nutrients to the environment for organisms to use.

## Tubifex worm

**Genus:** *Tubifex*

**Species:** Varies

**Size:** Up to 4 cm long

**Range:** Temperate freshwater ponds and streams

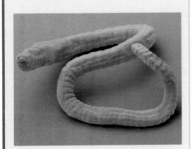

**Natural History:** Tubifex worms live on the bottom of ponds with their heads stuck into the substrate and tails waving in the water.

**Food:** Bacteria and detritus

**Predators:** Fish, amphibians, and crustaceans

**Shelter:** Burrows into the soil and gravel at the bottom of ponds

**Reproduction:** Tubifex worms are hermaphroditic (have both male and female sex organs).

**Abiotic Interaction:** Tubifex worms are found in the mud of many natural waterways.

**Biotic Interaction:** Tubifex worms are also raised for live food for tropical aquarium fish.

## Land snail

**Genus:** *Helix*

**Species:** Varies

**Size:** Shell up to 3 cm diameter

**Range:** Worldwide

**Natural History:** Snails have two sets of tentacles on their heads. The upper set contains nerve cells that are sensitive to light and smell. The two smaller tentacles on the bottom are sensitive to touch and can detect food, other snails, and surfaces. Snails are ectothermic, meaning they get body heat from their surrounding environment, and become inactive if too cool.

**Food:** Detritus, plants, and calcium sources

**Predators:** Birds, skunks, and raccoons

**Shelter:** Snails live on the ground and will move to cool, damp places to escape dry conditions. They retreat inside their shells and form a crusty layer over the opening to preserve the moisture in their bodies. This semi-hibernation is called estivation.

**Reproduction:** Snails are hermaphroditic (have both male and female sex organs). During mating, two snails exchange sperm; both may lay eggs. Eggs are laid underground.

**Abiotic Interaction:** Absence of calcium in the environment causes weak shells.

**Biotic Interaction:** Snails are often considered a garden and agricultural pest. For that reason, the US Department of Agriculture controls the movement of *Helix* across state lines. *Helix* is quarantined from some states. Some species of snail are eaten as a delicacy.

## Elodea

**Genus:** *Elodea*

**Species:** *canadensis*

**Size:** Sprigs up to 1 m long

**Range:** Throughout North America

**Natural History:** Elodea grows in freshwater ponds and slow-moving streams throughout North America. It is a member of the tape-grass family.

**Food:** Photosynthesis

**Predators:** Fish, insects, aquatic snails, crayfish, turtles, and salamanders

**Shelter:** Freshwater ponds and quiet streams

**Reproduction:** Elodea usually reproduces vegetatively (a form of asexual reproduction). It may produce small flowers at the tip, with seeds in small green capsules.

**Abiotic Interaction:** Productivity depends on light, water, and temperature levels.

**Biotic Interaction:** Elodea is a popular aquatic plant for home aquariums. Populations that are introduced to the wild can reproduce quickly and clog natural waterways. Sale of elodea is banned in some states because plants from aquariums have been released into local waterways and caused problems with overgrowth.

## Duckweed

**Genus:** *Lemna*

**Species:** *minor*

**Size:** 0.5 cm diameter

**Range:** Temperate freshwater ponds and lakes worldwide

**Natural History:** Duckweed is tiny, but huge populations can cover the entire surface of ponds and lakes. It is the smallest known flowering plant and a member of the duckweed family.

**Food:** Photosynthesis

**Predators:** Fish, aquatic snails, birds, water rats, and turtles

**Shelter:** Surface of freshwater ponds and lakes

**Reproduction:** Mostly by vegetative reproduction; sometimes sexual reproduction, if flowers appear

**Abiotic Interaction:** Productivity depends on light, water, and temperature levels. A population of duckweed can quickly cover the surface of calm waters.

**Biotic Interaction:** Duckweed is sold in aquarium stores for home aquariums. Duckweed can be eaten by many small fish in aquariums.

## Alfalfa

**Genus:** *Medicago*

**Species:** *sativa*

**Size:** 30–90 cm tall

**Range:** Temperate grasslands

**Natural History:** Alfalfa has a high tolerance for drought, cold, and heat. It has a taproot that can grow as deep as 15 m. It is a member of the pea family.

**Food:** Photosynthesis

**Predators:** Insects, birds, rodents, deer, and grazing livestock

**Shelter:** Mature plants reach a height of 1 m and can cover fields, with many insects living in the plants.

**Reproduction:** Small bunches of flowers at the ends of the stems develop into coiled seedpods. Flowers are pollinated by insects, primarily bees.

**Abiotic Interaction:** Productivity depends on light, water, and temperature levels. Growing alfalfa improves soil quality, so it is grown in alternating years with other food crops.

**Biotic Interaction:** Alfalfa is cultivated as a pasture crop or for hay.

## Rye grass

**Genus:** *Lolium*

**Species:** Varies

**Size:** 1–2 m tall

**Range:** Temperate grasslands of Europe, Asia, and North America

**Natural History:** Rye grass is a genus of tufted grass. This grass is native to Europe, Asia, and Africa, but is now grown worldwide. It is a member of the grass family.

**Food:** Photosynthesis

**Predators:** Birds, insects, rodents, deer, and grazing livestock

**Shelter:** Rye grass can cover fields. Left uncut, some species can grow nearly 1 m tall.

**Reproduction:** Flowers are wind-pollinated.

**Abiotic Interaction:** Productivity depends on light, water, and temperature levels.

**Biotic Interaction:** Rye grass is an important food for grazing livestock and is used for ornamental lawns. Rye grass should not be confused with rye, which is a different plant and an important grain cultivated throughout the world.

## Wheat

**Genus:** *Triticum*

**Species:** Varies

**Size:** 30–90 cm tall

**Range:** Grasslands worldwide

**Natural History:** Wheat grows best in a temperate climate with rainfall between 30 and 90 cm per year. It is a member of the grass family.

**Food:** Photosynthesis

**Predators:** Birds, insects, rodents, deer, humans, and grazing livestock

**Shelter:** Wheat can cover fields and is usually harvested when the plants are 1–2 m tall.

**Reproduction:** Flowers are wind-pollinated.

**Abiotic Interaction:** Productivity depends on light, water, and temperature levels.

**Biotic Interaction:** Wheat is an important grain, distributed throughout the world by humans. More land is devoted to growing wheat than to any other grain crop.

# Milkweed-Bug Hatching Investigation

A class of middle-school students decided to conduct some controlled experiments to find out what variables affect the hatching of milkweed-bug eggs. They planned experiments that they thought would help them understand milkweed-bug egg hatching, gathered the data, and organized it for others to share. They did not have time to summarize the results of the experiments or draw conclusions from those results. Here is the first part of their report.

**Title.** What Variables Might Affect Milkweed-Bug Egg Hatching?

**Purpose.** All organisms have limits on their populations. One limit on a population of milkweed bugs might be egg hatching. We decided to test three variables to see how they affect (1) the number of eggs that hatch and (2) the length of time before eggs hatch. The three variables we tested are temperature, humidity, and exposure to light.

**Experimental design.** We started with a large habitat of breeding milkweed bugs. One day before the experiments were scheduled to start, we put fresh pieces of polyester wool into the habitat. The next day we had several thousand new eggs to use in our experiment.

We used special equipment to control the variables for the experiments.

    A. A *temperature-control device* let us maintain precise temperatures.

    B. A *humidity-control device* let us maintain precise humidity (moisture in the air).

    C. A *light-control device* let us maintain precise exposure to light.

The standard hatching environment was 25°C, 50 percent humidity, and 12 hours of light exposure each day.

100 milkweed-bug eggs were placed in each experimental setting.

- In the temperature experiment, only temperature was changed. Humidity (50 percent) and light exposure (12 hours each day) were controlled.

- In the humidity experiment, only humidity was changed. Temperature (25°C) and light exposure (12 hours each day) were controlled.

- In the light experiment, only light exposure was changed. Humidity (50 percent) and temperature (25°C) were controlled.

Every 5 days the eggs were observed, and the number of eggs that had hatched was recorded. Nymphs were removed to another habitat, and the unhatched eggs were returned to the experimental conditions. The experiments continued for 30 days.

## Data

### Effect of Temperature on Milkweed-Bug Egg Hatching

A

Elapsed time in days

| Temperature (°C) | 0 days | 5 days | 10 days | 15 days | 20 days | 25 days | 30 days |
|---|---|---|---|---|---|---|---|
| 0° | 0 | 0 | 0 | 0 | 0 | 0 | 0 |
| 5° | 0 | 0 | 0 | 0 | 0 | 0 | 0 |
| 10° | 0 | 0 | 0 | 10 | 23 | 26 | 28 |
| 20° | 0 | 11 | 86 | 91 | 91 | 91 | 91 |
| 30° | 0 | 36 | 94 | 95 | 95 | 95 | 95 |
| 40° | 0 | 57 | 92 | 92 | 92 | 92 | 92 |
| 50° | 0 | 0 | 0 | 0 | 0 | 0 | 0 |
| 60° | 0 | 0 | 0 | 0 | 0 | 0 | 0 |

### Effect of Humidity on Milkweed-Bug Egg Hatching

B

Elapsed time in days

| Humidity (%) | 0 days | 5 days | 10 days | 15 days | 20 days | 25 days | 30 days |
|---|---|---|---|---|---|---|---|
| 0% | 0 | 26 | 80 | 96 | 96 | 96 | 96 |
| 25% | 0 | 22 | 88 | 91 | 91 | 91 | 91 |
| 50% | 0 | 28 | 90 | 95 | 95 | 95 | 95 |
| 75% | 0 | 26 | 86 | 95 | 95 | 95 | 95 |
| 100% | 0 | 21 | 87 | 96 | 96 | 96 | 96 |

### Effect of Light Exposure on Milkweed-Bug Egg Hatching

C

Elapsed time in days

| Light (hours per day) | 0 days | 5 days | 10 days | 15 days | 20 days | 25 days | 30 days |
|---|---|---|---|---|---|---|---|
| 0 | 0 | 28 | 88 | 94 | 94 | 94 | 94 |
| 6 | 0 | 22 | 83 | 94 | 94 | 94 | 94 |
| 12 | 0 | 25 | 90 | 97 | 97 | 97 | 97 |
| 18 | 0 | 23 | 82 | 91 | 91 | 91 | 91 |
| 24 | 0 | 26 | 88 | 96 | 96 | 96 | 96 |

**Results.** Look at the results for each experiment: temperature, humidity, and light exposure. Which of these conditions could be considered limiting factors? Use data to support your claims.

**Conclusions.** What do the results suggest about ways that milkweed-bug populations are limited in nature?

# Algae and Brine Shrimp Experiments

**Purpose.** Lab experiments were set up to determine if the abiotic factors of *light* and *temperature* limit population growth of algae and brine shrimp.

**Experimental design.** Four populations of planktonic algae and four populations of brine shrimp were placed in controlled environments. Population sizes were measured once each month for a year.

**Algae experiments.** Four identical aquariums were set up. Each had the same amount of Mono Lake water, nutrients (including carbon dioxide), and a small starter population of algae.

Two aquariums were maintained at constant temperatures (one at low temperature and one at high temperature), and the light was varied (changed).

The other two aquariums were maintained with constant light (one 9 hours a day and one 14 hours a day), and temperature was varied (changed).

**Algae Experimental Setup**

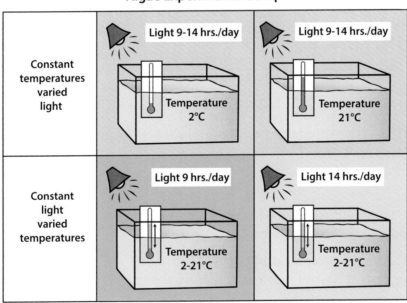

**Brine shrimp experiments.** Four identical aquariums were set up. Each had the same amount of Mono Lake water, food (powdered algae), and 1.0 g of brine shrimp eggs.

Two aquariums were maintained at constant temperatures (one at low temperature and one at high temperature), and the light was varied (changed).

The other two aquariums were maintained with constant light (one 9 hours a day and one 14 hours a day), and temperature was varied (changed).

### Brine Shrimp Experimental Setup

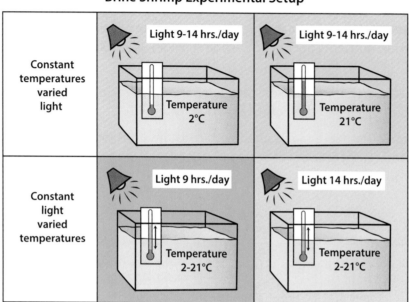

**Experimental procedure.** After they were set up, the eight aquariums were allowed to develop for 1 year.

## Light:

- The low-light aquariums received 9 hours of light each day. Nine hours of light represents the shortest days of the year at Mono Lake.

- The high-light aquariums received 14 hours of light each day. Fourteen hours of light represents the longest days of the year at Mono Lake.

- The variable-light aquariums received the amount of light each day that corresponds to the length of the calendar day at Mono Lake—9 hours in January, gradually increasing to 14 hours in June and July, then declining back to 9 hours in December.

## Temperature:

- The low-temperature aquariums were 2°C all year. Two degrees is the lowest temperature of Mono Lake in the winter.

- The high-temperature aquariums were 22°C all year. Twenty-two degrees is the highest temperature of Mono Lake in the summer.

- The variable-temperature aquariums started out cold (2°C) in January, warmed gradually to 22°C in July and August, and then cooled to 2°C by December.

Populations were checked once at the end of every month. A sample of aquarium water was removed and tested to see how much chlorophyll a was present. The amount of chlorophyll a, reported in micrograms per milliliter (µg/mL), is an indicator of the size of the algae population.

Populations of brine shrimp were counted by placing a sample of aquarium water under a microscope and counting all the shrimp of any size (larvae, juvenile, and adult). The result was converted to the number of brine shrimp in thousands per cubic meter (thousands/m³) of water.

## Results

### Planktonic algae experiments (Algae population in µg/mL)

| Temperature | Light (hrs/day) | Jan | Feb | Mar | Apr | May | Jun | Jul | Aug | Sep | Oct | Nov | Dec |
|---|---|---|---|---|---|---|---|---|---|---|---|---|---|
| 2°C | 9-14 | 1 | 3 | 27 | 68 | 86 | 91 | 92 | 96 | 94 | 92 | 93 | 95 |
| 21°C | 9-14 | 1 | 5 | 33 | 88 | 90 | 91 | 94 | 97 | 96 | 97 | 98 | 94 |
| 2 to 21°C | 9 | 1 | 2 | 10 | 31 | 55 | 82 | 92 | 96 | 94 | 89 | 92 | 93 |
| 2 to 21°C | 14 | 1 | 3 | 19 | 43 | 86 | 91 | 94 | 97 | 96 | 97 | 98 | 94 |

### Brine shrimp experiments (Brine shrimp in thousands/m³)

| Temperature | Light (hrs/day) | Jan | Feb | Mar | Apr | May | Jun | Jul | Aug | Sep | Oct | Nov | Dec |
|---|---|---|---|---|---|---|---|---|---|---|---|---|---|
| 2°C | 9-14 | 0 | 0 | 0 | 0 | 0 | 0 | 0 | 0 | 0 | 0 | 0 | 0 |
| 21°C | 9-14 | 3 | 40 | 54 | 58 | 57 | 54 | 53 | 45 | 53 | 68 | 70 | 74 |
| 2 to 21°C | 9 | 0 | 0 | 2 | 25 | 55 | 51 | 49 | 48 | 22 | 9 | 1 | 0 |
| 2 to 21°C | 14 | 0 | 0 | 3 | 22 | 56 | 55 | 51 | 48 | 25 | 8 | 2 | 0 |

## Conclusions

1. Based on the experimental results, which factors placed limits on the algae populations? What is your evidence?

2. Based on the experimental results, which factors placed limits on the brine shrimp populations? What is your evidence?

3. What additional abiotic and biotic factors might limit population size in Mono Lake?

# Mono Lake Data

**Background.** Because of its unique ecology, Mono Lake has been an interesting place for scientists to study. Good scientific study involves accurate data recording. A lot is known about the organisms that live in the lake and the abiotic conditions that affect the organisms in the ecosystem.

These three pages have graphs that show how some of the populations in the Mono Lake ecosystem change over the course of a year and how the abiotic factors change over the course of a year.

Study the graphs. Look for relationships (1) between populations of different species and (2) between organisms and abiotic factors in the ecosystem.

## Abiotic data

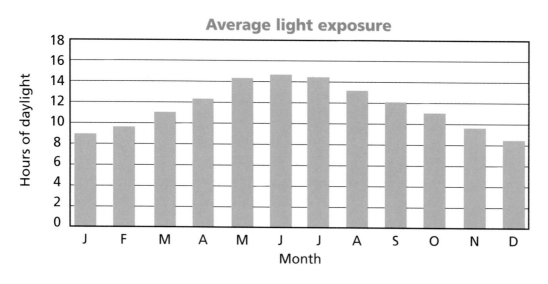

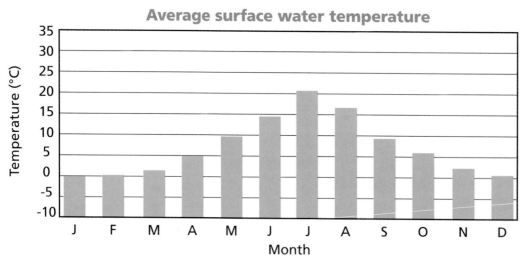

# Population data

## Average planktonic algae population

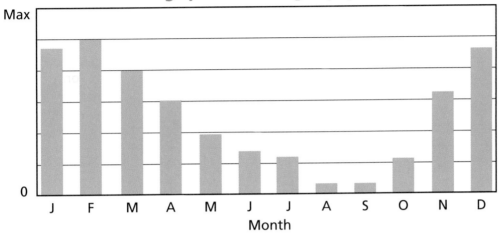

## Average brine fly population

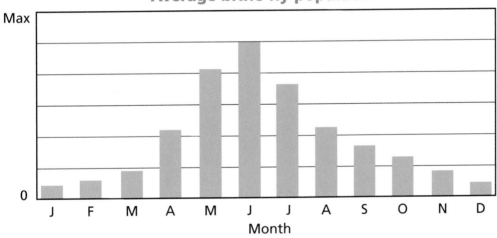

## Average brine shrimp population

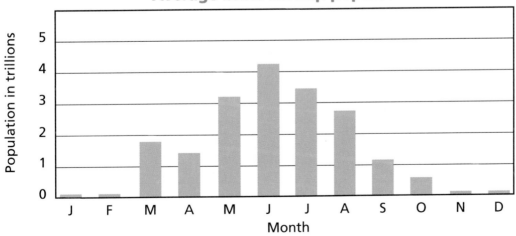

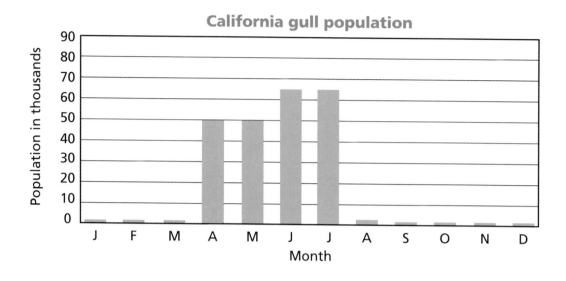

**California gull population**

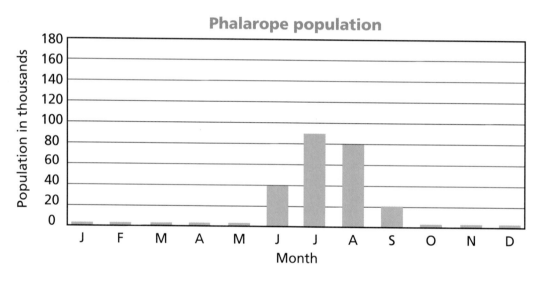

**Phalarope population**

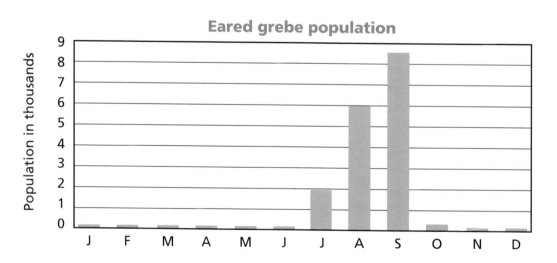

**Eared grebe population**

## Conclusions

1. When does the planktonic algae population peak? When does the brine shrimp population peak? What explanation do you have for the timing of each peak?

2. What is the relationship between water temperature and the brine shrimp and brine flies in the Mono Lake ecosystem?

3. When do the birds arrive?

4. Discuss the similarities and differences in the population graphs of the birds (gulls, phalaropes, and grebes).

5. What is the relationship between the birds and the other organisms?

6. What do you think is going on with the populations at Mono Lake in April?

# Science Safety Rules

**1** Always follow the safety procedures outlined by your teacher. Follow directions, and ask questions if you're unsure of what to do.

**2** Never put any material in your mouth. Do not taste any material or chemical unless your teacher specifically tells you to do so.

**3** Do not smell any unknown material. If your teacher asks you to smell a material, wave a hand over it to bring the scent toward your nose.

**4** Avoid touching your face, mouth, ears, eyes, or nose while working with chemicals, plants, or animals. Tell your teacher if you have any allergies.

**5** Always wash your hands with soap and warm water immediately after using chemicals (including common chemicals, such as salt and dyes) and handling natural materials or organisms.

**6** Do not mix unknown chemicals just to see what might happen.

**7** Always wear safety goggles when working with liquids, chemicals, and sharp or pointed tools. Tell your teacher if you wear contact lenses.

**8** Clean up spills immediately. Report all spills, accidents, and injuries to your teacher.

**9** Treat animals with respect, caution, and consideration.

**10** Never use the mirror of a microscope to reflect direct sunlight. The bright light can cause permanent eye damage.

# Glossary

**abiotic** nonliving

**aerobic cellular respiration** a process by which organisms convert sugar into usable energy

**aquatic** of the water

**atmosphere** the thin layer of gases around Earth that extends about 600 kilometers above the surface

**atom** a particle that is the basic building block of matter

**autotroph** an organism that makes its own food

**bioaccumulation** the process by which toxin accumulates in bodies of predators

**biodiversity** the variety of life

**biodiversity index** a measure of biodiversity for an area by dividing the total number of species by the total number of individuals

**biomass** the mass of matter produced by organisms in an ecosystem

**biome** collection of ecosystems that have similar environments and organisms

**biosphere** the sum total of all living organisms

**biotic** consisting of living organisms and products of organisms

**calorie** a unit used to measure heat energy

**carbohydrate** food in the form of sugar or starch

**carnivore** an organism that eats other animals

**carrying capacity** the maximum size of a population that a given environment can support

**chlorophyll** a green pigment in chloroplasts that captures light energy to make sugars during photosynthesis

**clutch** a cluster of eggs

**community** all the interacting populations in a specified area

**consumer** an organism that eats other organisms

**control** to conduct an experimental test to compare results with tests where a variable was changed

**controlled experiment** an experiment in which the observer is able to standardize all but one variable to measure results

**cultural** in the context of ecosystem services, benefits of an ecosystem such as recreation and tourism

**decomposer** an organism that consumes parts of dead organisms and converts all the biomass into simple chemicals

**dehydration** when the body doesn't have enough water to function normally

**detritivore** an organism that eats detritus, breaking the organic material into smaller parts

**detritus** dead organic material

**ecosystem** a system of interacting organisms and nonliving factors in a specified area

**ecosystem engineer** a keystone species that changes the ecosystem it inhabits

**ecosystem service** a benefit that humans obtain from the environment

**energy** the capacity to do work. Most energy used by organisms comes from the Sun.

**environment** the surroundings of an organism, including the living and nonliving factors

**exoskeleton** a tough, outer covering that insects and other organisms have for protection

**extinct** no longer existing

**extinction** when the last of a species' population dies out and it can no longer reproduce

**food** a substance that provides energy and nutrients for organisms

**food chain** a sequence of organisms that eat one another in an ecosystem

**food web** all the feeding relationships in an ecosystem

**habitat** a place where an organism lives that supports its requirements for life

**herbivore** an organism that eats only plants

**heterotroph** an organism that cannot make its own food and must eat other organisms

**hydrosphere** all the water on Earth including the ocean, lakes, rivers, streams, aquifers, polar icecaps, glaciers, snowpacks, permafrost, and condensation

**incomplete metamorphosis** the process of gradual maturing of an insect through stages (egg, nymphal stages or instars, adult)

**individual** a single organism

**inference** an explanation or assumption that people make based on their knowledge, experiences, or opinions

**instar** an immature nymphal stage of an insect as it grows into an adult form

**interdependent** dependent on each other

**introduced species** a species that is brought to a place by humans and becomes part of the local ecosystem

**invasive species** an introduced species that has a negative impact on an ecosystem

**keystone species** a species that is critical for the overall health of an ecosystem

**kilocalorie** a unit equal to 1,000 calories

**limiting factor** any biotic or abiotic component of the ecosystem that controls the size of a population

**lithosphere** the rocky, mineral part of Earth

**mass** the amount of matter in something

**migrate** to move from one area to another according to seasonal changes

**molecule** a particle made of two or more smaller particles held together by chemical bonds

**molting** the process of shedding an exoskeleton in order to grow

**native species** a kind of organism that has been part of an ecosystem for a long time

**nymph** an immature bug

**observation** noticing the properties of an object or event with one or more of the five senses (sight, hearing, touch, smell, and taste)

**observational study** an experiment in which the observer collects data over time without interacting with the area of study

**omnivore** a consumer that eats both plants and animals

**organism** a living thing

**photosynthesis** the process by which producers make energy-rich molecules (food) from water and carbon dioxide in the presence of light

**phytoplankton** a huge array of photosynthetic microorganisms that are free-floating in water

**polar zone** the climate zone that is closest to the North and South Poles (latitude 60°–90° north and latitude 60°–90° south)

**population** all the individuals of one kind (one species) in a specified area at one time

**population study** an experiment in which the observer collects data over time for one population in an ecosystem

**predator** an organism, usually an animal, that eats other organisms

**prey** an organism, usually an animal, that is eaten by another organism

**primary consumer** organisms that eat producers; also called first-level consumers

**proboscis** a tube-like beak for sucking fluids from plants. True bugs have this structure.

**producer** an organism that is able to produce its own food through photosynthesis

**provisioning** in the context of ecosystem services, benefits of an ecosystem such as food, water, or fuel

**regulating** in the context of ecosystem services, benefits of an ecosystem such as climate regulation or disease control

**reproductive potential** the theoretical unlimited growth of a population over time

**sampling** a technique to count the organisms in a selected area to make inferences about the total number of organisms

**secondary consumer** organisms that eat primary consumers; also called second-level consumers

**species** a kind of organism. Members of a species are all the same kind of organism and are different from all other kinds of organisms.

**supporting** in the context of ecosystem services, benefits of an ecosystem that support other services, such as water cycling and soil formation

**sustain** to last or maintain over a long period of time

**sustainable** can last or maintain itself over a long period of time

**temperate zone** the climate zone lying between the Tropic of Cancer and the Arctic Circle in the Northern Hemisphere or between the Tropic of Capricorn and the Antarctic Circle in the Southern Hemisphere

**terrestrial** of the land

**tertiary consumer** organisms that eat secondary consumers; also called third-level consumers

**trophic level** a functioning role in a feeding relationship through which energy flows

**trophic pyramid** a trophic-level diagram in which the largest layer at the base is the producers; the first-level, second-level, and third-level consumers are in the layers above

**tropical zone** the climate zone closest to the equator (latitude 30° south to 30° north)

**tufa tower** a naturally occurring, gray, lumpy structure that forms underwater in a salt lake because of a chemical reaction between calcium and salt

**variation** the range of expression of a trait within a population

**zooplankton** microscopic adult animals and larval forms of animals found free-floating in fresh water and sea water

# Index

## A
**abiotic**, 10, 24, 27, 67–69, 107
**aerobic cellular respiration**, 52, 107
**aquatic**, 14, 24, 75, 107
**Arctic National Wildlife Refuge**, 13
**Arctic tundra**, 13
**atmosphere**, 30, 107
**atom**, 44, 107
**autotroph**, 51, 62, 107

## B
**bacteria**, 63–64
**bioaccumulation**, 56, 107
**biodiversity**, 74–78, 107
**biodiversity index**, 75, 107
**biomass**, 51, 60, 66, 107
**biome**, 11–24, 75, 107
**biosphere**, 23, 31, 107
**Biosphere 2**, 31–36
**biotic**, 23, 24, 66–67, 107

## C
**calorie**, 43, 107
**carbohydrate**, 48, 107
**carbon dioxide**, 35–38, 45, 51, 52, 69, 70–71, 72
**carnivore**, 59, 107
**carrying capacity**, 69–71, 107
**Carson, Rachel**, 55
**chlorophyll**, 51, 107
**climate**, 23
**climate change**, 69, 77
**clutch**, 7, 107
**community**, 9–10, 107
**competitor**, 81–82
**conservation**, 78, 87–88
**consumer**, 51, 62, 63, 107
**control**, 32, 107
**controlled experiment**, 32, 107
**coral reef**, 17, 37–38
**cultural**, 13, 107

## D
**decomposer**, 59, 61, 62, 63–64, 107
**dehydration**, 42, 107
**Delaware Water Gap National Recreation Area**, 14
**detritivore**, 64, 107
**detritus**, 64, 107

## E
**ecoscenario**, 11–24, 107
**ecosystem**, 10, 11, 25, 52, 61, 69, 75, 107
**ecosystem engineer**, 76, 107
**ecosystem service**, 12, 108
**El Yunque National Forest**, 15
**energy**, 30, 33, 36, 41, 43, 44, 45–49, 50–52, 62, 67, 69, 70, 108
**energy transfer**, 60
**environment**, 10, 55–57, 67–68, 108
**Environmental Protection Agency**, 57
**Everglades National Park**, 16

## F
**exoskeleton**, 6, 108
**extinct**, 76, 108
**extinction**, 68, 76, 108

**Florida Keys National Marine Sanctuary**, 17
**food**, 5, 44–49, 62, 70, 76, 108
**food chain**, 56, 75, 78, 108
**food web**, 26, 28, 59, 69, 70–71, 75, 108
**fungi**, 63–64

## G
**Gaines, David**, 86

## H
**habitat**, 3, 76, 108
**herbivore**, 59, 60, 108
**heterotroph**, 51, 62, 108
**hydrosphere**, 31, 108

## I
**incomplete metamorphosis**, 6–7, 108
**individual**, 4, 108
**inference**, 3–4, 108
**instar**, 6, 108
**interdependent**, 75, 108
**introduced species**, 79, 108
**invasive species**, 77, 79–85, 108

## K
**keystone species**, 75, 108
**kilocalorie**, 43, 108

## L
**limiting factor**, 64–71, 108
**lithosphere**, 30, 108

## M
**Maathai, Wangari**, 53–54
**mass**, 51, 60, 108
**mating habits**, 7–8, 69
**migrate**, 26, 29, 69, 73, 108
**milkweed bugs**, 5–8
**molecule**, 44, 108
**molting**, 6, 108
**Mono Lake**, 25–29, 66, 67, 69, 72–73, 85–87
**Monongahela National Forest**, 18
**Monterey Bay National Marine Sanctuary**, 19, 75

## N
**native species**, 79, 108
**nymph**, 6, 108

## O
**observation**, 3–4, 108
**observational study**, 32, 108
**omnivore**, 59, 66, 108
**organism**, 8, 9–10, 11–12, 24, 28, 31, 33–35, 44, 45–47, 48, 50–51, 62–63, 108

## P
**pesticide**, 55–57
**photosynthesis**, 17, 19, 37, 46, 50–51, 52, 63, 68, 69, 70, 108
**phytoplankton**, 17, 47, 109
**polar zone**, 23, 109
**pollution**, 57, 77
**population**, 7–8, 9, 66, 72, 75, 109
**population study**, 32, 109
**predation**, 66, 73
**predator**, 60–61, 75, 80–81, 109
**prey**, 60–61, 109
**primary consumer**, 58, 59, 61, 70, 109
**proboscis**, 5, 109
**producer**, 50–51, 61, 62, 63, 70–71, 109
**provisioning**, 12, 109

## R
**regulating**, 12, 109
**reproduction**, 7–8, 70
**reproductive environment**, 67–68
**reproductive potential**, 65, 109

## S
**safety rules**, 106
**sampling**, 75, 109
**season**, 68–69, 72–73
**secondary consumer**, 58, 61, 70, 109
**Sonoran Desert National Monument**, 20
**species**, 8, 109
**sunlight**, 33, 37, 45–46, 48, 51, 69, 70
**supporting**, 12, 109
**sustain**, 33, 70, 109
**sustainable**, 70, 109

## T
**Tallgrass Prairie National Preserve**, 21
**temperate zone**, 23, 109
**terrestrial**, 14, 24, 75, 109
**tertiary consumer**, 58, 61, 109
**trophic level**, 58–62, 109
**trophic pyramid**, 60–61, 109
**tropical rain forest**, 15, 24, 37
**tropical zone**, 23, 109
**tufa tower**, 26, 109

## V
**variation**, 24

## W
**water**, 41–43, 52, 85–87

## Y
**Yellowstone National Park**, 22, 75

## Z
**zooplankton**, 47, 61, 109